Advances in Edible Oil Processing

Advances in Edible Oil Processing

Editors

Suzana Ferreira-Dias
Xuebing Xu
Fátima Peres

MDPI • Basel • Beijing • Wuhan • Barcelona • Belgrade • Manchester • Tokyo • Cluj • Tianjin

Editors

Suzana Ferreira-Dias
Universidade de Lisboa
Portugal

Xuebing Xu
Wilmar Global Research and Development Center
China

Fátima Peres
Instituto Politécnico de Castelo Branco, Escola Superior Agrária
Portugal

Editorial Office
MDPI
St. Alban-Anlage 66
4052 Basel, Switzerland

This is a reprint of articles from the Special Issue published online in the open access journal *Life* (ISSN 2075-1729) (available at: https://www.mdpi.com/journal/life/special_issues/edible_oil).

For citation purposes, cite each article independently as indicated on the article page online and as indicated below:

LastName, A.A.; LastName, B.B.; LastName, C.C. Article Title. *Journal Name* **Year**, *Volume Number*, Page Range.

ISBN 978-3-0365-3467-1 (Hbk)
ISBN 978-3-0365-3468-8 (PDF)

Contents

About the Editors

Suzana Ferreira-Dias is Associate Professor with Habilitation at the School of Agriculture (Instituto Superior de Agronomia, ISA) at the University of Lisbon (Universidade de Lisboa, ULisboa), Lisbon, Portugal. Her main research is centered in the fields of Food Engineering and Bioenergy, focused on the technology/biotechnology of oils and fats. She has been working mainly on (i) biocatalysis in non-conventional media to produce structured lipids, emulsifiers and flavors for the food and pharmaceutical industries as well as biodiesel from non-edible oils, (ii) enzyme-catalyzed production of glucose and prebiotics from lignocellulosic materials, (iii) extraction optimization and characterization of vegetable oils and virgin olive oil, (iv) enrichment of edible oils with bioactive compounds, and (v) sensory analysis of foods for quality control or product development.

Xuebing Xu is Chief Scientist of Wilmar China/Yihai Kerry. He received his PhD from the Technical University of Denmark. He was earlier Professor in Agro-Biotechnology Science at Aarhus University in Denmark and is currently Adjunct Professor at two universities. His research covers lipid technology, enzyme technology, structured lipids, food functionality, and food safety, among others.

Fátima Peres is Professor at the School of Agriculture (Escola Superior Agrária, ESACB) of the Polytechnic Institute of Castelo Branco (Instituto Politécnico de Castelo Branco, IPCB), Castelo Branco, Portugal and is also Head of the Laboratory Food Technology and Analytical Instrumentation at ESACB. Her main research is centered in olive oil technology and food chemistry and analysis.

Preface to "Advances in Edible Oil Processing"

Edible oil production has been a traditional industry over the last century. The processing technology, however, has advanced step by step, especially with the general progress of science and technology. The requirements for edible processing are also changing. Besides quality and cost, the consideration of sustainability and safety in all aspects has been in great demand. In the recent decades, a few subjects on technology development have been hot topics of research and development in the edible oil processing industry. Bioprocessing using enzymatic processes or fermentation technology has been widely applied for edible oil processing. Besides quality improvements and the possibility to produce functional lipids, biotechnology offers a great advantage of green and sustainable processing, including points such as less wastewater pollution, less air pollution, and more energy efficiency. The development of moderate or mild processing is also being pursued. The target is maintenance of more nutritional components while limiting negative components as much as possible. For traditional products, technology upgrades are needed to meet the modern needs of sustainability and increase resource utilization. For the development of new products, various names arise, such as structured lipids, oleogels, flavored edible oils, etc., in which processing is a central concern.

Thus, this Special Issue "Advances in Edible Oil Processing" is a collection of six manuscripts covering several hot topics on this theme. Three manuscripts focused on biocatalysis applied to (i) oil processing, namely the enzymatic degumming of rice bran oil, using different commercial phospholipases, as alternative to the conventional acid degumming (Rodrigues et al.), and the production of novel lipids with improved functional properties by lipase-catalyzed reactions, such as (ii) low-calorie structured lipids (SL) rich in linoleic acid, which is an essential fatty acid for humans, using argan oil (Simões et al.), and also (iii) human milk fat substitutes (HMFSs), addressed in the review of Jian et al. Moreover, the use of milk fat globule membrane together with HMFs, as supplements in infant formulas, as well as the legislation about HMFs, are also discussed in this review.

Novel products with improved biological values were also developed, namely (i) a gourmet flavored oil enriched in phenolic compounds, which was obtained by co-extraction of overripe olives with low aroma potential with a thyme from the Mediterranean region (*Thymus mastichina* L.) (Peres et al.) and (ii) a sauce prepared with blends of olive oil and soybean oil and supplemented with olive leaf polyphenol extracts obtained by ultrasound-assisted extraction, which improved the oxidative stability of the sauce (Aliyari and Rezaei). Especially in Asia, flavor-enhanced edible oils, rich in volatile pyrazines and formed by roasting or thermally processing through Maillard reactions, are highly appreciated by consumers. Xu et al. developed an efficient and accurate analytical methodology to quantify pyrazines in flavored oils based on multiple headspace solid phase microextraction-arrow–gas chromatography mass spectrometry (MHS-SPME-arrow–GCMS) to quantify 13 pyrazines in flavor-enhanced edible oils (e.g., peanut oil, sesame oil, and rapeseed oil). In conclusion, these six publications in this Special Issue cover important topics in edible oil processing.

We thank Mrs. Beryl Liu from *Life* for her initial invitation and proposal for the organization of this Special Issue. We also thank the journal office for efficient processing of papers for publication. We express special thanks to the authors of the six papers for their support and prompt handling during the publication process.

Suzana Ferreira-Dias, Xuebing Xu, Fátima Peres
Editors

life

Article

Enzymatic Degumming of Rice Bran Oil Using Different Commercial Phospholipases and Their Cocktails

Mayara S. Rodrigues [1], Rafaela M. Dos Passos [1], Paula V. de A. Pontes [1], Marcela C. Ferreira [2], Antonio J. A. Meirelles [1], Christian V. Stevens [3], Guilherme J. Maximo [1] and Klicia A. Sampaio [1,*]

1 Department of Food Engineering, Faculty of Food Engineering, University of Campinas, Campinas 13083-862, Brazil; mayararodrigues4181@hotmail.com (M.S.R.); r211223@dac.unicamp.br (R.M.D.P.); paulapontes15@gmail.com (P.V.d.A.P.); tomze@fea.unicamp.br (A.J.A.M.); maximo@unicamp.br (G.J.M.)

2 Faculty of Technology, University of Campinas, Limeira 13484-332, Brazil; marcela@ft.unicamp.br

3 Department of Green Chemistry and Technology, Faculty of Bioscience Engineering, Ghent University, 9000 Ghent, Belgium; Chris.Stevens@UGent.be

* Correspondence: klicias@unicamp.br

Abstract: Rice bran oil is a highly nutritious vegetable oil, as it is rich in tocols and γ-oryzanol. Degumming is the first step in the vegetable oil refining process, and its main objective is the removal of phospholipids or gums. In the present study, enzymatic degumming trials were performed on crude rice bran oil using the phospholipases PLA1, Purifine® PLC, their mixture (PLA1/PLC), and a cocktail known as Purifine® 3G. Enzymatic degumming applying 50 mg/kg of PLA1 for 120 min resulted in a residual phosphorus content of 10.4 mg/kg and an absolute free fatty acid increase of 0.30%. Enzymatic degumming applying 300 mg/kg of Purifine® PLC for 120 min at 60 °C resulted in a residual phosphorus content of 67 mg/kg and an absolute diacylglycerol increase of 0.41%. The mixture of phospholipases and the cocktail presented approximately 5 mg/kg of residual phosphorus content after the reaction times. For all degumming processes, the preservation of minor components such as tocols and γ-oryzanol were observed. These results indicate that the use of enzyme mixtures or their cocktails to attain low phosphorus content and high diacylglycerol/free fatty acid conversion during enzymatic degumming is a viable alternative.

Keywords: vegetable oils; phospholipase; enzymatic degumming; phospholipids

Citation: Rodrigues, M.S.; Dos Passos, R.M.; Pontes, P.V.d.A.; Ferreira, M.C.; Meirelles, A.J.A.; Stevens, C.V.; Maximo, G.J.; Sampaio, K.A. Enzymatic Degumming of Rice Bran Oil Using Different Commercial Phospholipases and Their Cocktails. *Life* **2021**, *11*, 1197. https://doi.org/10.3390/life11111197

Academic Editors: Suzana Ferreira-Dias, Xuebing Xu and Fátima Peres

Received: 24 September 2021
Accepted: 28 October 2021
Published: 6 November 2021

1. Introduction

Rice bran oil (RBO) is a type of cooking oil obtained from rice bran (*Oryza Sativa* L.) through appropriate technological procedures [1]. The oil contains in its composition saponifiable and unsaponifiable fractions. Moreover, from a nutritional perspective, RBO is exceptionally rich in minor components, such as tocotrienols, phytosterols, and γ-oryzanol, where, if combined, exert a wide spectrum of biological activities [2].

In order to reach the edible condition, crude oil obtained as a result of the extraction process must be submitted to a refining process, whose objective is to remove impurities and undesirable compounds [3]. Degumming is the first step in the refining of vegetable oils, and its main goal is the removal of phospholipids or gums. The main phospholipids present in rice oil are: phosphatidylcholine (PC), phosphatidylinositol (PI), phosphatidylethanolamine (PE), and phosphatidic acid (PA) [4].

Phospholipids are classified according to their degree of hydration (hydratable and non-hydratable). Hydratable phospholipids (HPL) become insoluble in oil in the presence of water and are easily separated by centrifugation. Most non-hydratable phospholipids (NHPL) are complexed with calcium (Ca), magnesium (Mg), and iron (Fe) salts, and to be removed, they need the addition of a chelating agent (citric acid or EDTA) to sequester metal ions, allowing their precipitation and separation by centrifugation [5,6].

Enzymatic degumming is a process for the removal of phospholipids from crude oil using phospholipases (enzymes). The phospholipases hydrolyze the ester bonds present on the phospholipid molecules, and diacylglycerols (DAGs) or free fatty acids (FFA) are released [7]. The most commonly used phospholipases are: phospholipase A1 and phospholipase A2 that remove the fatty acid from position 1 and 2 with respect to glycerol, and phospholipase C (PLC) that hydrolyzes the bond between the acylglycerol and the phosphate group to release diacylglycerols (DAG) [3,7,8]. Besides phosphorus removal, the use of phospholipases has the advantage of oil yield increase, effluent generation decrease, and production costs decrease [4,7].

A new enzyme cocktail has been developed, Purifine® 3G; the mixture consists of Purifine® PLC, PLA2, phosphatidylinositol-specific phospholipase C (PI-PLC) that acts on the phosphatidylcholine (PC), phosphatidylethanolamine (PE), and phosphatidylinositol (PI) [9]. The enzymatic cocktail has a high efficient conversion of phospholipids (PLs) into mostly diglycerides (DAGs), phosphates, some free fatty acids (FFAs), and some lysophospholipids (LPLs) [10]. Therefore, the cocktail produces an FFA and DAG increase and phosphorus decrease, which are advantages over using only a single enzyme.

The use of enzymatic degumming increases the oil yield and brings advantages to the oil industry. In addition, the association of enzymatic degumming with other processes, such as the production of biodiesel, allows for the use of unrefined oil, making processing cheaper [11]. Thus, it is necessary to study the best process parameters and oil types for expanding industrial application. Therefore, the main objective of the present study is to evaluate process parameters such as enzyme concentration and reaction time, applied to RBO using Lecitase Ultra (PLA1), Purifine® PLC, and Purifine® 3G. The experimental results were evaluated concerning the residual phosphorus content (P-content), the FFA increase, the DAG increase, and the content of tocols and γ-oryzanol of the degummed RBO.

2. Materials and Methods

2.1. Raw Material and Reagents

The crude rice bran oil was kindly donated by IRGOVEL (Pelotas-RS, Brazil). All chemicals used are either ultra-performance liquid chromatography (UPLC) or analytical grade. Sodium hydroxide (NaOH) and citric acid (CA) were purchased from Sigma Aldrich (São Paulo, Brazil). The diolein standard (purity $\geq$ 99%), the tetradecane, and the derivatizing agent (BSTFA) were purchased from Sigma Aldrich (São Paulo, Brazil).

2.2. Enzymes

The phospholipase C type Purifine® PLC was donated by DSM Company (Delft, The Netherlands) with an activity of 22,000 PLCU/g. The cocktail Purifine® 3G (PLC + PI-PLC + PLA2) was donated by DSM Food Specialties (Delft, The Netherlands) with an activity of 16,900 PLCU/g. The phospholipase A type Lecitase Ultra (PLA1) was donated by Novozymes (The Netherlands) with an activity of 10 KLU/g.

2.3. Physicochemical Analysis

The free fatty acid (FFA) content was determined by titration according to the AOCS official method Ca 5a-40 [12] and was expressed as % by weight of oleic acid. The fatty acid profile of crude rice bran oil was analyzed by gas chromatography (GC), according to the AOCS official method Ce 1–62 [13]. Phosphorus content was measured by inductively coupled plasma (ICP) according to AOCS official method Ca 20–99 [14]. The pH was measured directly with a pH electrode in the gums fraction.

The acylglycerol composition was measured according to the AOCS official method Cd 11b-91 [15]. Approximately 0.05 g of the oil samples was dissolved in 100 µL of tetradecane and 300 µL of derivatizing agent (BSTFA). The mixture was heated at 70 °C for 20 min. Then, 50 µL of derivatized sample was transferred to vials and diluted with 1 mL of hexane and injected in a gas chromatography (Agilent Technologies 7890A, Santa Clara, CA, USA, using GC Agilent 7890A, with OnColumn injection and DB-5HT capillary column

(15 m × 0.32 mm i.d. ×0.10 μm film thickness). The diacylglycerols were identified using a diolein standard.

2.4. Nuclear Magnetic Resonance (NMR) Analysis

The analysis of phospholipid composition was measured by Nuclear Magnetic Resonance (NMR) employing a Bruker Avance III 400 MHz automatic spectrometer. Triphenyl phosphate was used as internal standard [16].

2.5. Analysis of Minor Components

The γ-oryzanol content was determined according to the Codex Alimentarius methodology [1], which uses spectroscopy, in which n-heptane is used as a solvent. First, a scan of the γ-oryzanol in heptane solution was carried out over the entire range of the UV-visible spectrum to determine the wavelength at which maximum absorption occurs. A calibration curve was, then, constructed with solutions of known concentrations (0.030–0.20 mg/mL) of γ-oryzanol in heptane at the maximum absorption wavelength. The determination of γ-oryzanol in crude rice bran oil was carried out by weighing approximately 0.02 g of the sample in a 25 mL volumetric flask and diluting with heptane. Then, the solution was read with 315 nm absorbance.

The determination of tocols content was carried out according to the methodology of Ansolin et al. [17]. The oil was diluted in isopropanol to a concentration of approximately 8000 $\mu g.mL^{-1}$. The samples were filtered through hydrophobic PTFE filters and, then, followed for analysis. For liquid chromatography analysis, was utilized a Waters Chromatographic system composed of an ultra-performance liquid chromatography (UPLC) model Acquity, coupled with a mass spectrometer. The separation of the compounds was performed on a UPLC BEH C18 column (1.7 μm, 2.1 mm × 100 mm) operating at 30 °C. The injection volume was 5 μL. All samples were analyzed in triplicate.

2.6. Water Degumming (WDG)

Water degumming was performed in order to verify the best percentage of water to be added. For the performance of the process, the crude oil was initially heated at 80 °C and water percentages of 3, 5, 7, and 10% (*w*/*w*)—relative to the oil mass—were added, and the mixture was homogenized with mechanical stirring (350 rpm) for 15 min and, then, centrifuged (10,000 rpm/15 min) for the separation of degummed oil from gum.

2.7. Chemical Conditioning (CC)

The chemical conditioning aimed to adjust the pH value for maximal enzyme activity. In this case, crude RBO was heated to 80–85 °C, and citric acid was added as a 30% aqueous solution. The oil followed high shear mixing (1 min/16,000 rpm), and then, the mixture was stirred for 15 min/350 rpm. Then, a 14% NaOH aqueous solution was added and followed a stirring period (1 min/16,000 rpm). After, the gums and the oil were separated by centrifugation (15 min/1000 rpm), and both were sent for analysis.

2.8. Enzymatic Degumming Experiments (PLA1, Purifine® PLC, Purifine® 3G, and Combinations)

The enzymatic degumming experiments were performed with 400 g of crude RBO. The first step of the enzymatic degumming process using PLA1, Purifine® PLC, and Purifine® 3G was carried out similarly to the steps of the chemical conditioning (CC) in order to adjust the pH, however, without the centrifugation step. The oil was conditioned for 15 min at 80 °C with stirring at 350 rpm. After conditioning, the temperature of the oil mixture was reduced to 52–60 °C, depending on the type of the enzyme. Then, a certain amount of water (3%, relative to the weight of the oil) and a predefined amount of PLA1 (10–70 mg/kg), Purifine® PLC (100–400 mg/kg), the combination PLA1/PLC-1G (50–300 mg/kg), or Purifine® 3G (300 mg/kg) were added. For PLA1 and PLC experiments, first, the ideal concentrations of the enzymes were found, and then, the reaction time was analyzed. The mixture was homogenized under high shear (16,000 rpm) for 1 min to disperse the enzyme in the oil/water emulsion. After, the oil mixture was kept at the required temperature under stirring (350 rpm) for a period of time (0–120 min). The degumming reaction was stopped by heating the mixture for 15 min at 85 °C. Subsequently, the oil mixture underwent centrifugation (15 min/1000 rpm) to separate the degummed oil from the gums.

2.9. Statistical Analysis

All measurements were performed in triplicate with all data expressed as mean value ± standard deviation of independent experiments in triplicate. Statistical analysis was performed with STATISTICA 7.0. The differences among the means were determined by the Tukey test. Significant differences were declared at $p \leq 0.05$.

3. Results

The fatty acid composition, free fatty acid content, acylglycerol composition, and minor components such as tocols and γ-oryzanol content of crude rice bran oil are listed in Table 1. As expected, rice bran oil is mostly composed of TAG, but important amounts of acylglycerols were also detected. The free fatty acids, which are final degradation products of TAGs, represent about 5% of the crude oil. Rice bran oil contains oleic acid as the predominant fatty acid, followed by linoleic and palmitic acid, with minor amounts of additional fatty acids. Similar results were obtained by Lüdtke [18], where the levels of fatty acids in greater proportion in rice bran were: oleic (32.8–35.6%), linoleic (30.8–33.5%), and palmitic (19.5–21.1%).

Table 1. Characterization of crude rice bran oil.

Parameter	RBO [1]
TAG (%)	92.73 ± 0.03
DAG (%)	2.17 ± 0.04
MAG (%)	0.13 ± 0.00
FFA (%)	4.97 ± 0.02
Fatty Acid Composition	**(%)**
Miristic (C14:0)	0.28 ± 0.02
Palmitic (C16:0)	20.20 ± 0.01
Estearic (C18:0)	1.75 ± 0.10
Oleic (C18:1)	39.56 ± 0.05
Linoleic (C18:2)	36.37 ± 0.80
Linolenic (C18:3)	1.84 ± 0.02
Minor Components	**(mg/kg)**
α-tocopherol	146.96 ± 0.12
β/γ-tocopherol	149.70 ± 0.28
α-tocotrienol	38.22 ± 0.54
β/γ-tocotrienol	461.45 ± 2.89
Total	796.33 ± 26.8
γ-oryzanol (%)	1.80 ± 0.03

[1] RBO: rice bran oil. Content is the mean of triplicate ± S.D.

According to the results, β/γ-tocotrienol presented the highest concentration in crude rice bran oil followed by β/γ-tocopherol and α-tocopherol. Tocopherols and tocotrienols are important antioxidants present in rice bran oil, which are related to the prevention of coronary disease, cataract formation, and lowering the levels of plasma triacylglycerols and cholesterol [19]. The γ-oryzanol content was analyzed, and the crude rice bran oil presented a content of 1.8%. Similar results were obtained by Van Hoed et al. [20] when evaluating the influence of chemical refining on the major and minor compounds of rice bran oil. According to Wang et al. [21], γ-oryzanol is efficient in serum normalization of total cholesterol, triglycerides, and LDL-cholesterol, and it induces FFA level reductions and high-density lipoprotein (HDL) cholesterol increases.

Table 2 shows the minerals content and the phospholipids composition present in crude rice bran oil. The total phosphorus content, determined by the inductively coupled plasma (ICP) at crude rice bran oil, is at a level of 426 mg/kg, while the phospholipids content evaluated by P-NMR reached 0.91%. The most relevant minerals for the oil refining process are P, Ca, Mg, and Fe. Residual phosphorus (>10 mg/kg) can cause oil darkening, resulting in off-flavors and imposing difficulties on the oil downstream processing [22]. In addition, the amounts of Ca, Fe, and Mg are directly related to non-hydratable phospholipids and, consequently, to the facility of the degumming process.

Table 2. Minerals content and phospholipids composition in the crude rice bran oil.

Parameter	RBO [1]
Minerals (mg/kg)	
P	426.0 ± 6.0
Ca	9.60 ± 0.06
Fe	5.05 ± 0.08
Mg	60.0 ± 0.20
Phospholipids (%)	
PC	0.39 ± 0.04
PE	0.25 ± 0.02
PI	0.20 ± 0.01
PA	0.07 ± 0.03
Total (%)	0.91 ± 0.03

[1] RBO: rice bran oil. Results of minerals content. Phospholipids composition. Contents are the mean of triplicate.

Rice bran oil was also analyzed concerning its phospholipid composition, and according to the results, phosphatidylcholine (PC), together with phosphatidylethanolamine (PE), represents 0.64% of the oil that contains a total of 0.91% of phospholipids. The oil also contains additional amounts of phosphatidylinositol (PI) and phosphatidic acid (PA), which can also be degraded by phospholipases.

3.1. Water Degumming (WDG)

The water degumming process has the main goal of removing the so-called hydratable phospholipids. The process was carried out using water proportions ranging from 3 to 10% *w*/*w*, relative to the oil mass. Figure 1 shows the minerals content found for the different proportions of water. According to Lamas, Constenla, and Raab [23], the phospholipids remaining after the water treatment can be considered as non-hydratable phospholipids (NHPL).

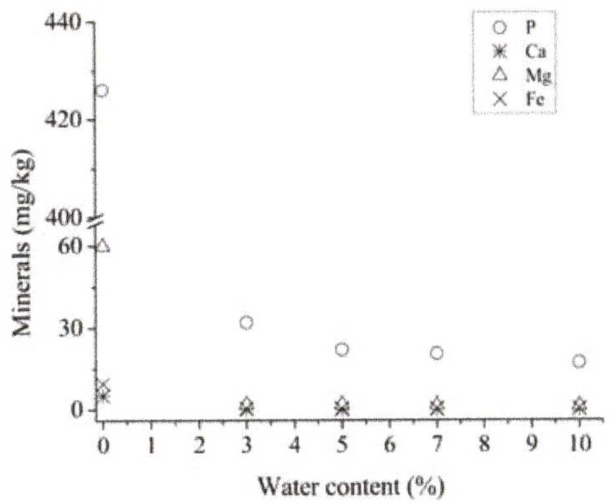

Figure 1. Minerals content after water degumming of crude rice bran oil.

As can be seen, there was a reduction in the content of all minerals present in crude rice bran oil after water degumming. The phosphorus content in the crude oil was reduced from 426 mg/kg to 32.0, 22.0, 20.5, and 17.2 mg/kg when adding 3, 5, 7, and 10% water, respectively. However, there was only a small reduction in the phosphorus content (32.0–17.2 mg/kg) when the water content was increased from 3 to 10%, not showing great advantages for the process, besides increasing the costs of process and the consequent generation of effluents. Thus, 3% of water would already be a suitable value for carrying out the aqueous degumming process.

3.2. *Enzymatic Degumming with PLA1*

Enzymatic degumming is used for removing phospholipids or gums, and the process is carried out using phospholipases. The phospholipase A1 (PLA1) is used industrially, and due to its action in the sn1 position of the phospholipid, this enzyme produces free fatty acids and lysophospholipids. Fatty acids from the hydrolysis of phospholipids participate in increasing the oil yield, while lysophospholipids are removed by centrifugation, constituting the gums.

In general, a chemical conditioning using citric acid and sodium hydroxide is required to improve the hydratability of the non-hydratable phospholipids and to achieve the optimal pH value for the enzyme. When the chemicals were added to the vegetable oil, the content of phosphorus in the crude rice bran oil decreased from 426 mg/kg to 88 mg/kg. Similar results were obtained by Sampaio et al. [8] and Jiang et al. [24] when studying the degumming of crude corn and soybean oil.

The PLA1 (Lecitase Ultra) experiments were performed with the addition of different enzyme concentrations to crude rice bran oil in order to monitor the changes in the residual phosphorus and FFA content. Figure 2 shows the phosphorus content and the FFA content as a function of the enzyme concentration within a reaction time of 120 min. As can be observed, the phosphorus content decreases, and the FFA content increases as the enzyme concentration increases. For enzyme concentrations varying from 10 to 50 mg/kg, the phosphorus content was reduced from 22.8 to 10.4 mg/kg, respectively. When the enzyme dosage was increased to 70 mg/kg, the residual phosphorus content was reduced only to 9.5 mg/kg. Regarding the FFA content, the evaluation of the FFA increase is also based on the amount of free fatty acids (4.97%) present in the degummed oil after the chemical conditioning. The enzyme Lecitase Ultra (PLA1) used in this work acts selectively in the sn1 position of the phospholipid, releasing fatty acids. Theoretically, conversion of 0.1% of phospholipids gives a formation of 0.036% FFA [25]. Therefore, as the crude rice bran oil has 0.91% of phospholipids, the generation of 0.33% of acidity would be expected. With an increase in the PLA1 dosage from 10 to 50 mg/kg, the absolute FFA increases linearly from 0.12% to 0.30%. However, a further increase in the enzyme concentration (70 mg/kg) results in a very poor FFA increase (0.31%), possibly due to the complete consumption of the substrate (gums). Considering these results, and the fact that 10 mg/kg is the maximum value to the degummed oil follow the physical refining [26], the enzyme dosage was set as standard at 50 mg/kg.

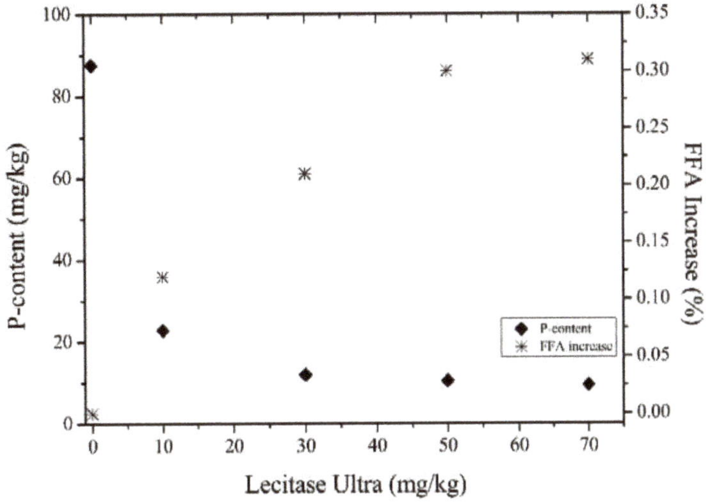

Figure 2. Effect of enzyme (PLA1) concentration on the residual phosphorus content and absolute FFA increase (reaction conditions: 120 min; 52 °C; 3% water; pH 5.1).

Figure 3 shows the residual phosphorus content and the absolute FFA content obtained after the degumming process of crude rice bran oil with the enzyme Lecitase Ultra (PLA1) for different reaction times. By increasing the reaction time from 15 to 60 min, the content of residual phosphorus decreases from 21 to 15 mg/kg, while the absolute FFA content increases from 80.11% to 0.24%. A further increase in the reaction time from 90 to 120 min resulted in a phosphorus decrease from 13 to 10 mg/kg, and an absolute FFA increase varying from 0.26 to 0.30%. Therefore, 120 min of reaction time was fixed as an optimum value for the reaction time of crude rice bran oil.

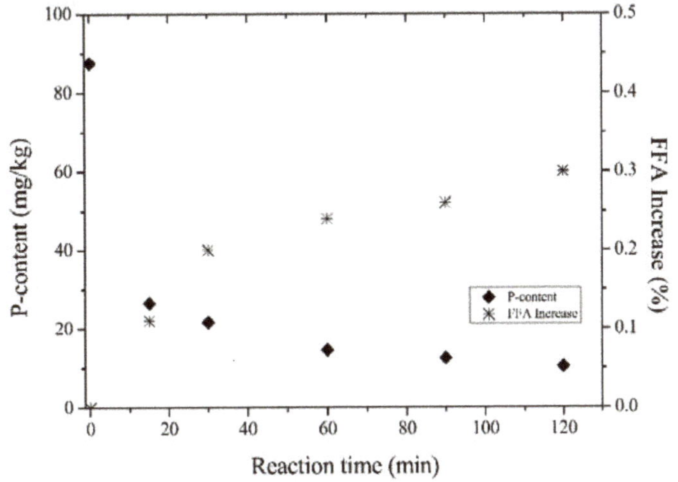

Figure 3. Effect of the enzyme (PLA1) reaction time on the residual phosphorus content and absolute FFA increase in the degummed oil (reaction conditions: 50 mg/kg; 52 °C; 3% water; pH 5.1).

Jahani et al. [27] studied the enzymatic degumming of rice bran oil using the enzyme Lecitase Ultra obtained from an experimental sample of Thermomyces lanuginosus/Fusarium oxysporum, with a concentration of 50 mg/kg. According to the authors, a reduction in the phosphorus content to values <10 mg/kg was achieved after a reaction time of 240 min. This result is probably due to the use of an older enzyme generation.

3.3. Enzymatic Degumming with Purifine PLC

The enzyme Purifine® PLC is a phospholipase that acts selectively on the phospholipids PC and PE. The phospholipase C hydrolyzes the bond between the acylglycerol and the phosphate group realizing diacylglycerols (DAG) that are recognized as part of the oil, and hence contributing to the oil yield. Figure 4 shows the residual phosphorus content and the absolute DAG increase as a function of different Purifine PLC dosage. It is noticeable that for enzyme concentrations varying from 100 to 200 mg/kg, the residual phosphorus content remains practically constant. When enzyme concentrations varying from 300 to 400 mg/kg are added, the residual phosphorus content is reduced to values of 68 and 47 mg/kg, respectively. According to Gupta [9], depending on the degummed oil and the process employed, the P-content can reach values greater than 100 mg/kg.

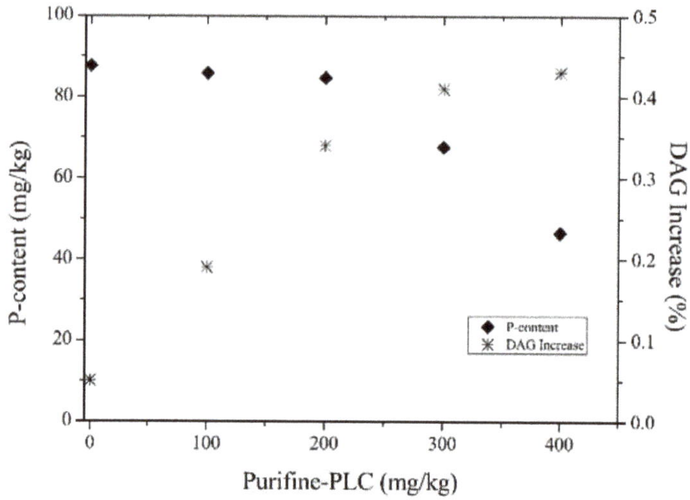

Figure 4. Effect of Purifine® PLC dosage on the residual phosphorus content and absolute DAG increase (reaction conditions: 120 min; 60 °C; 3% water; pH 6.0).

Concerning the relative DAG increase, its evaluation was also carried out considering that the Purifine PLC enzyme acts specifically towards PC and PE, which results in a total of 0.65% (Table 2). According to Dayton and Galhardo [28], the amount of DAG that may be generated can be calculated using the original PC + PE content of the oil multiplied by the ratio between the molecular weight of the diolein and the average molecular weights of the phospholipids (605/750). An observed reaction efficiency of approximately 85% is used, and it is explained experimentally by the complete reaction of PC, but only partial reaction of PE, due to the relative rate of hydration. Thus, for a phospholipid content of 0.65%, a total DAG increase of 0.45% would be expected. The DAG content varied from 2.22% for the chemical conditioned oil, up to a level of 2.63% when using a concentration of 300 mg/kg of enzyme. Therefore, after degumming, the absolute DAG increase was 0.41%. Further increase in the enzyme concentration to 400 mg/kg resulted in only a 3% DAG increase. Considering the obtained results, it was established that 300 mg/kg is the most suitable enzyme concentration for the degumming of rice bran oil.

After defining the best concentration of Purifine® PLC to carry out the experiments applied to crude rice bran oil, the reaction kinetics was evaluated. Thus, enzymatic degumming was performed with Purifine® PLC for a concentration of 300 mg/kg at different reaction times. Figure 5 shows the effect of the reaction time on the absolute

DAG increase and residual phosphorus content of degummed rice bran oil. By adding 300 mg/kg of enzyme and increasing the reaction time from 15 to 60 min, it is possible to see that the residual phosphorus content decreased from 85 to 75 mg/kg, while the absolute DAG content increased from 0.04 to 0.20%. With an increase in the reaction time from 90 to 120 min, the residual phosphorus content remained practically constant (68–67 mg/kg), while the relative DAG increase varied from 0.32–0.41%. Hence, 120 min of reaction time was fixed as an optimum value for Purifine PLC reaction for crude rice bran oil.

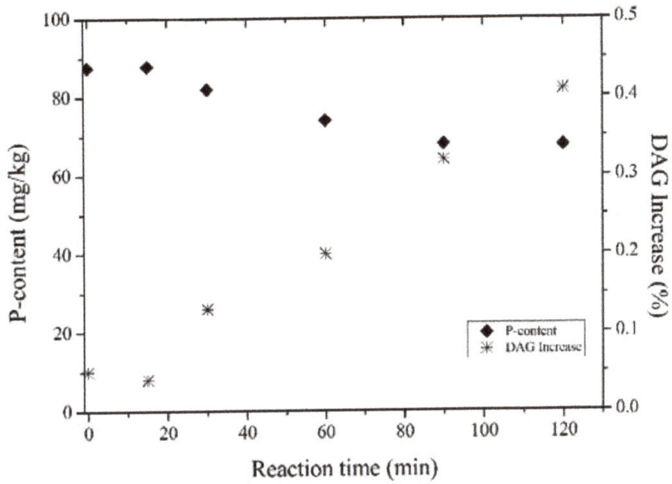

Figure 5. Effect of enzyme (PLC) reaction time on the residual phosphorus content and absolute DAG increase (reaction conditions: 300 mg/kg Purifine® PLC; 60 °C; 3% water; pH 6.0).

Sampaio et al. [8] studied the enzymatic degumming of corn oil using 200 mg/kg of the Purifine® PLC enzyme and found that, at 120 min of reaction time, the relative increase in the DAG content recached 0.54%, while the P-content was of 27 mg/kg. The authors also found that an increase in the reaction time to 240 min contributed only to a slight increase in DAG content and a decrease in the residual P, not justifying the adoption of a longer reaction time.

3.4. *Enzymatic Degumming with Purifine® 3G and Combinations (Purifine PLC/PLA1)*

As Purifine PLC is an enzyme that acts selectively towards PC and PE, the obtained degummed oil has a phosphorus content comparable with water degummed oil, with phosphorus levels varying from 50 to 100 mg/kg. On the one hand, the enzyme Lecitase Ultra (PLA1) degrades all the phospholipids, but it produces FFA as a neutral oil, and higher FFA content in the degummed oil requires a larger fatty acid stripper, and it also promotes small additional oil loss when the FFA content increases. A possible solution to the mentioned issues can be the use of a mixture of PLC and PLA or other enzyme cocktails such as Purifine 3G added to the crude oil. Purifine® 3G cocktail is a set of phospholipases that acts on all phospholipids (PA, PI, PC, and PE), as it consists of a mixture of the enzymes Purifine® PLC, PLA2, and PI-PLC [9]. Therefore, it acts not only by increasing the oil yield (DAG and FFA), but also by reducing the minerals content. It is worth mentioning that within the literature, there are only few works on Purifine® 3G enzyme until the present moment [10].

The addition of the enzyme mixture (PLC/PLA) and the enzyme cocktail (Purifine 3G) to crude rice bran oil was performed in both cases for two different reaction times, 60 min and 120 min. For the dosage of the enzyme mixture, we chose the optimum concentrations found previously (PLA1:50 mg/kg and PLC: 300 mg/kg), and in the case of the cocktail (Purifine 3G), the supplier data (300 mg/kg) were used. The obtained degummed oil was evaluated taking into account the DAG and FFA content and also the residual phosphorus content.

Table 3 shows the residual phosphorus content, the DAG and FFA content of the degummed rice bran oil after the action of the enzyme mixture, and the cocktail for reaction times of 60 min and 120 min. As can be seen, the enzyme mixture as well as the cocktail were effective in reducing the residual phosphorus content to values lower than 5 mg/kg for both reaction times. These results indicate that the use of either enzyme for 60 min of reaction time is already sufficient for the degumming of rice bran oil under the conditions employed. Regarding the oil yield increase, which was evaluated by the quantification of FFA and DAG content in the degummed rice bran oil, it can be seen that a longer reaction time resulted in practically constant values for the FFA content; however, it presented a higher DAG content, probably due to a high conversion extent. It is important to highlight that this behavior was observed for the enzyme mixture as well as for the cocktail.

Table 3. DAG content, FFA content, and P-content of degummed rice bran oil for enzyme mixture and cocktail at different reaction times.

Combinations and Cocktail	Reaction Time (min)	DAG (%)	FFA (%)	P-Content (mg/kg)
CC	0	2.22 ± 0.19 ^{a}	4.99 ± 0.14 ^{a}	87.6 ± 0.04 ^{a}
PLC-1G/PLA$_1$	60	2.52 ± 0.02 ^{a}	5.24 ± 0.02 ^{a}	5.22 ± 0.02 ^{b}
PLC-1G/PLA$_1$	120	2.64 ± 0.06 ^{a}	5.29 ± 0.00 ^{a}	5.26 ± 0.01 ^{b}
PLC-3G	60	2.62 ± 0.05 ^{a}	5.21 ± 0.00 ^{a}	5.00 ± 0.50 ^{b}
PLC-3G	120	2.76 ± 0.15 ^{a}	5.27 ± 0.04 ^{a}	5.01 ± 0.12 ^{b}

Chemical conditioning (CC). $^{a,\,b}$: Different letters in the same column indicate significant differences ($p < 0.05$).

Jiang et al. [4] studied enzymatic degumming on eight different varieties of vegetable oils using a mixture of enzymes PLC/PLA1 for 3.5 h of reaction time (0.5 h used for acid pretreatment and 3 h for enzymatic degumming), and the authors found a residual phosphorus content lower than 5 mg/kg for all oils, therefore, indicating the effectiveness of the use of the enzyme mixture added to crude vegetable oils.

3.5. *Minor Components (Tocols and γ-Oryzanol)*

The minor components present in rice bran oil, which correspond to the constituents of vitamin E (tocopherols and tocotrienols) and γ-oryzanol, constitute a great advantage to the vegetable oil and were evaluated after the degumming treatments. Table 4 presents the tocol content after water degumming and enzymatic degumming using PLA1 and Purifine® PLC for different water concentrations and reaction times, respectively. As can be seen, among the tocols, β/γ-tocotrienol presented the highest value in the degummed oil, followed by β/γ-tocopherol, α-tocopherol, and α-tocotrienol. δ-tocopherol content was not detected in the oil. Similar results were found by Lüdtke [18] when studying the water degumming of crude rice bran oil.

Table 4. Content of tocols after water degumming and enzymatic degumming using PLA_1 and Purifine® PLC.

Water Content (%)	β/γ-Tocotrienol	α-Tocotrienol	β/γ-Tocopherol	α-Tocopherol	Total (mg/kg)
0*	461.45 ± 2.89 [a]	38.22 ± 0.54 [a]	149.70 ± 0.28 [a]	146.96 ± 0.12 [a]	796.33 ± 18.13 [a]
3	460.11 ± 2.80 [a]	19.22 ± 2.80 [b]	120.21 ± 1.81 [b,c]	133.23 ± 1.10 [a]	732.77 ± 1.84 [b]
10	438.46 ± 2.92 [a,b]	15.02 ± 1.16 [b]	116.15 ± 2.12 [c]	124.00 ± 2.30 [b]	693.63 ± 15.3 [b]
PLA_1 (min)	**β/γ-Tocotrienol**	**α-Tocotrienol**	**β/γ-Tocoferol**	**α-Tocopherol**	**Total (mg/kg)**
0	458.30 ± 1.26 [a]	29.18 ± 0.02 [a,b]	136.11 ± 0.01 [b]	140.23 ± 0.04 [a]	763.82 ± 21.16 [a,b]
30	408.70 ± 2.13 [b,c]	17.54 ± 0.62 [b]	148.06 ± 0.83 [a]	145.28 ± 1.28 [a,b]	719.58 ± 14.3 [b]
60	390.56 ± 3.80 [b,c]	16.54 ± 0.23 [b]	152.56 ± 1.24 [a]	143.43 ± 0.45 [a,b]	703.09 ± 2.64 [b]
120	408.42 ± 0.31 [b,c]	16.02 ± 0.07 [b]	137.53 ± 0.45 [a]	121.62 ± 0.86 [b]	683.59 ± 11.1 [b]
Purifine PLC (min)	**β/γ-Tocotrienol**	**α-Tocotrienol**	**β/γ-Tocopherol**	**α-Tocopherol**	**Total (mg/kg)**
0	457.6 ± 1.20 [a]	28.17 ± 0.02 [a,b]	136.15 ± 0.01 [b]	141.23 ± 0.03 [a,b]	763.15 ± 20.10 [a,b]
30	408.70 ± 5.72 [b,c]	17.54 ± 0.43 [b]	148.06 ± 0.76 [a]	145.28 ± 1.54 [a,b]	719.58 ± 6.77 [b]
60	409.31 ± 3.40 [b,c]	17.12 ± 0.30 [b]	146.83 ± 3.05 [a]	144.15 ± 2.36 [a,b]	717.41 ± 5.24 [b]
120	420.69 ± 1.15 [a,b]	20.00 ± 0.16 [b]	124.42 ± 0.59 [b,c]	131.31 ± 0.48 [b]	696.42 ± 9.60 [b]

Crude rice bran oil (0*); chemical conditioning (0). [a, b, c]: Different letters in the same column indicate significant differences ($p < 0.05$).

Regarding the water degumming process, it is possible to verify that increasing the amount of water from 0 to 3% contributed to a significative reduction in tocol content when comparing to crude rice bran oil. The tocol losses reached approximately 9% for the highest amount of water added (10%), and these losses can be related to own process conditions such as high temperatures and oxidation reactions. When evaluating the enzymatic degumming processes using PLA1 and PLC for different reaction times, it can be observed that the chemical conditioning already promotes a reduction in tocol content, and it probably happens due to removal of part of the tocols during the degumming treatment that uses acids and base, besides the reaction time use for the process. Van Hoed et al. [29] verified that rice bran oil presented a relative change in the tocols' composition when evaluating the physical refining deacidification, and the authors attribute the changes to the evaporation process caused by the temperatures and the low pressures.

Table 4 shows the minor components in the rice bran oil that underwent degumming with PLA1 within a reaction time of 120 min. The loss of tocol content was of approximately 10.5% from 0 to 120 min. Only the total tocol content of water degumming presented a significant difference from crude rice bran oil, and all enzymatic degumming did not differ significantly from chemical conditioning. Van Hoed et al. [29] utilized a physical deacidification per column for rice bran oil that consisted of an adapted type of washing and found that the rice bran oil presented a change in the relative composition of the tocols.

Table 5 shows the tocol content after the use of the enzyme mixture and the cocktail. According to the results, it can be seen that the β/γ-tocotrienol presented the highest values, thus, remaining in accordance with the results for the other types of degumming previously performed without a mixture. One should also point out that Purifine® 3G cocktail showed significant β/γ-tocopherol differences for 60 and 120 min, and a similar difference was observed for β/γ-tocotrienol regarding the enzyme mixture, which can be related to losses associated to the processes.

Table 5. Tocol content after enzymatic degumming using combination (PLCPLA$_1$) and cocktail (Purifine® 3G).

Treatments	Time (min)	β/γ-Tocotrienol (mg/kg)	α-Tocotrienol (mg/kg)	β/γ-Tocopherol (mg/kg)	α-Tocopherol (mg/kg)	Total (mg/kg)
CC	0	458.30 ± 1.30 ^{a}	29.20 ± 0.50 ^{a}	136.10 ± 0.34 ^{a}	140.20 ± 0.28 ^{a}	763.8 ± 21.2 ^{a}
PLC/PLA1	60	436.76 ± 0.35 ^{a}	29.40 ± 0.70 ^{a}	151.85 ± 0.22 ^{a}	137.05 ± 0.23 ^{a}	755.06 ± 9.89 ^{a}
PLC/PLA1	120	404.18 ± 1.24 ^{b}	27.45 ± 0.80 ^{a}	152.86 ± 0.21 ^{a}	135.81 ± 0.30 ^{a}	720.30 ± 14.71 ^{a}
PLC-3G	60	452.14 ± 1.60 ^{a}	22.83 ± 1.30 ^{a}	129.88 ± 0.44 ^{b}	138.21 ± 0.38 ^{a}	743.06 ± 1.41 ^{a}
PLC-3G	120	433.47 ± 1.40 ^{a}	23.26 ± 3.20 ^{a}	132.46 ± 0.58 ^{b}	133.91 ± 0.46 ^{a}	723.10 ± 12.7 ^{a}

Chemical conditioning (CC). $^{a, b}$: Different letters in the same column indicate significant differences ($p < 0.05$).

The γ-oryzanol content represents an important compound present in rice bran oil, and it was measured in the present study after all the degumming processes. It was observed that the content of γ-oryzanol present in the oil after chemical conditioning was 1.7%, and the γ-oryzanol content remained in the range of 1.6–1.7% for all degumming processes, with no major changes. The present study is in accordance to Krishna [30] who was able to prove that the content of γ-oryzanol in the oil was not affected by degumming.

4. Conclusions

The present study showed that the use of PLA1, enzyme mixture (PLA1/Purifine PLC), and enzyme cocktail (Purifine 3G) was effective in reducing the residual phosphorus content of degummed rice bran oil to less than 10 mg/kg under the optimal conditions. However, rice bran oil degummed with Purifine PLC alone would require further treatment in order to be adequate for physical refining. After PLA1, about a 0.30% of absolute FFA increase was obtained. Degumming with PLC enzyme did not result in an FFA increase, but it increased the absolute DAG content (0.41%) by hydrolyzing PC ad PE. Treatments with enzyme mixtures (PLA1/PLC) and the cocktail (PLC 3G) resulted in both FFA and DAG increases in levels varying between 85 and 95% of the total phospholipid content. Concerning the minor components, which are represented in degummed rice bran oil by tocols and γ-oryzanol, it was verified that the chemical conditioning promoted a significant decrease in tocol content, while the γ-oryzanol content remained practically unchanged when comparing with the crude oil. For the enzyme treatments at different reaction times, there was a significant reduction when comparing the tocols results with chemical conditioning; however, when evaluating for the different reaction times, the tocols' reduction was not always significant. Therefore, it can be concluded that enzymatic degumming with PLA1, Purifine® PLC, enzyme mixture (PLA1/PLC), and the cocktail (Purifine® 3G) is efficient in degumming rice bran oil and can be used industrially.

Author Contributions: M.S.R., writing—original draft preparation and conceptualization; R.M.D.P. and P.V.d.A.P., methodology and formal analysis; M.C.F., writing—original draft preparation; A.J.A.M., supervision, funding acquisition; C.V.S., writing—reviewing and editing; G.J.M., supervision, writing—reviewing; K.A.S., conceptualization, writing—reviewing and editing and funding acquisition. All authors have read and agreed to the published version of the manuscript.

Funding: The authors thank the funding agencies: CNPq (429873/2018-2; 307398/2019-6; 406963/2016-9; 132428/2018-0), FAPESP (2014/21252-0). This study was financed in part by the Coordenação de Aperfeiçoamento de Pessoal de Nível Superior, Brasil (CAPES), Finance Code 001.

Institutional Review Board Statement: Not applicable.

Informed Consent Statement: Not applicable.

Acknowledgments: The authors wish to express their sincere gratitude to Irgovel Indústria Riograndense de Óleos Vegetais for kindly supplying the samples of crude rice bran oil.

Conflicts of Interest: There are no conflict to declare.

References

1. Alimentarius, C. Codex Standard for Named Vegetable Oils. *Codex-Stan* **2015**, *210*, 1–13.
2. Lai, O.-M.; Jacoby, J.J.; Leong, W.-F.; Lai, W.-T. Nutritional Studies of Rice Bran Oil. *Rice Bran Rice Bran Oil* **2019**, *1*, 19–54. [CrossRef]
3. Sampaio, K.A.; Zyaykina, N.; Wozniak, B.; Tsukamoto, J.; De Greyt, W.; Stevens, C.V. Enzymatic Degumming: Degumming Efficiency versus Yield Increase. *Eur. J. Lipid Sci. Technol.* **2015**, *117*, 81–86. [CrossRef]
4. Jiang, X.; Chang, M.; Jin, Q.; Wang, X. Application of Phospholipase A1 and Phospholipase C in the Degumming Process of Different Kinds of Crude Oils. *Process Biochem.* **2015**, *50*, 432–437. [CrossRef]
5. Cowan, D.; Korsholm, N. Development in Enzymatic Degumming. In Proceedings of the Presentation at the 7th Euro Fed Lipid Congress, Graz, Austria, 18–21 October 2009.
6. Manjula, S.; Jose, A.; Divakar, S.; Subramanian, R. Degumming Rice Bran Oil Using Phospholipase-A1. *Eur. J. Lipid Sci. Technol.* **2011**, *113*, 658–664. [CrossRef]
7. Clausen, K. Enzymatic Oil-Degumming by a Novel Microbial Phospholipase. *Eur. J. Lipid Sci. Technol.* **2001**, *103*, 333–340. [CrossRef]
8. Sampaio, K.A.; Zyaykina, N.; Uitterhaegen, E.; De Greyt, W.; Verhé, R. LWT—Food Science and Technology Enzymatic Degumming of Corn Oil Using Phospholipase C from a Selected Strain of Pichia Pastoris. *LWT Food Sci. Technol.* **2019**, *107*, 145–150. [CrossRef]
9. Gupta, M.K. *Practical Guide to Vegetable Oil Processing*; Elsevier: Amsterdam, The Netherlands, 2017; pp. 60–76.
10. Nikolaeva, T.; Rietkerk, T.; Sein, A.; Dalgliesh, R.; Bouwman, W.G.; Velichko, E.; Tian, B.; Van As, H.; van Duynhoven, J. Impact of Water Degumming and Enzymatic Degumming on Gum Mesostructure Formation in Crude Soybean Oil. *Food Chem.* **2020**, *311*, 126017. [CrossRef]
11. Passos, R.; Ferreira, R.; Batrista, E.; Meirelles, A.J.A.; Maximo, G.J.; Ferreira, M.C.; Sampaio, K.A. Degumming Alternatives for Edible Oils and Biodiesel Production. *Food Public Health* **2019**, *9*, 139–147. [CrossRef]
12. American Oil Chemists' Society AOCS. *Official Method Ca 5a-40, Free Fatty Acids Official Methods and Recommended Pratices of the AOCS*, 6th ed.; AOCS Press: Urbana, IL, USA, 2012.
13. American Oil Chemists' Society AOCS. Official method Ce 1-62, acid composition by Gas Chromatography official Methods and recommended Practices of the AOCS. In *Champaign*; AOCS Press: Urbana, IL, USA, 2012.
14. American Oil Chemists' Society AOCS. *Official Method Ca 20-99, Inductively Coupled Plasma (ICP) Official Methods and Recommended Prectices of the AOCS*, 6th ed.; AOCS Press: Urbana, IL, USA, 2012.
15. American Oil Chemists' Society AOCS. *Official Method Cd 11b-91, Dyglycerides by Capillary Gas Chromatography Official Methods and Recommended Practices of the AOCS*, 6th ed.; AOCS Press: Urbana, IL, USA, 2012.
16. Diehl, B.W.K. NMR Spectroscopy of Natural Substances. *NMR Spectrosc. Pharm. Anal.* **2008**, *1*, 194–196. [CrossRef]
17. Ansolin, M.; Souza, P.T.; Meirelles, A.J.A.; Batista, E.A.C. Tocopherols and Tocotrienols: An Adapted Methodology by UHPLC/MS without Sample Pretreatment Steps. *Food Anal. Methods* **2017**, *10*, 2165–2174. [CrossRef]
18. Lüdtke, F.L. Estudo da degomagem e clarificação de óleo bruto do farelo de arroz (Oryza Sativa). *Visando Refino Físico* **2016**, *1*, 1–31.
19. Qureshi, A.A.; Khan, D.A.; Mahjabeen, W.; Qureshi, N. Dose-Dependent Modulation of Lipid Parameters, Cytokines and RNA by [Delta]-Tocotrienol in Hypercholesterolemic Subjects Restricted to AHA Step-1 Diet. *Br. J. Med. Med. Res.* **2015**, *6*, 351–366. [CrossRef]
20. Van Hoed, V.; Depaemelaere, G.; Ayala, J.V.; Santiwattana, P.; Verhé, R.; De Greyt, W. Influence of Chemical Refining on the Major and Minor Components of Rice Bran Oil. *J. Am. Oil Chem. Soc.* **2006**, *83*, 315–321. [CrossRef]
21. Wang, O.; Liu, J.; Cheng, Q.; Guo, X.; Wang, Y.; Zhao, L.; Zhou, F.; Ji, B. Effects of Ferulic Acid and –Oryzanol on High-Fat and High-Fructose Diet-Induced Metabolic Syndrome in Rats. *PLoS ONE* **2015**, *10*, e0118135. [CrossRef]
22. Dijkstra, A.J. Recent Developments in Edible Oil Processing. *Eur. J. Lipid Sci. Technol.* **2009**, *111*, 857–864. [CrossRef]
23. Lamas, D.L.; Constenla, D.T.; Raab, D. Effect of Degumming Process on Physicochemical Properties of Sunflower Oil. *Biocatal. Agric. Biotechnol.* **2016**, *6*, 138–143. [CrossRef]
24. Jiang, X.; Chang, M.; Wang, X.; Jin, Q.; Wang, X. A Comparative Study of Phospholipase A 1 and Phospholipase C on Soybean Oil Degumming. *J. Am. Oil Chem. Soc.* **2014**, *91*, 2125–2134. [CrossRef]
25. De Greyt, W. Current and Future Technologies for the Sustainable and Cost-Efficient Production of High Quality Food Oils. *Eur. J. Lipid Sci. Technol.* **2012**, *114*, 1126–1139. [CrossRef]
26. Yang, B.; Zhou, R.; Yang, J.G.; Wang, Y.H.; Wang, W.F. Insight into the Enzymatic Degumming Process of Soybean Oil. *J. Am. Oil Chem. Soc.* **2008**, *85*, 421–425. [CrossRef]
27. Jahani, M.; Alizadeh, M.; Pirozifard, M.; Qudsevali, A. Optimization of Enzymatic Degumming Process for Rice Bran Oil Using Response Surface Methodology. *LWT Food Sci. Technol.* **2008**, *41*, 1892–1898. [CrossRef]
28. Dayton, C.L.G.; Galhardo, F. Enzymatic Degumming Utilizing a Mixture of PLA and PLC Phospholipases. U.S. Patent 20080182322A1, 17 February 2015.
29. Van hoed, V. Quality Assessment of the High Value Vegetable Oils by Characterization of Minor Components. *Fac. Biosci. Eng. Ghent Univ. Belgium.* **2010**, *1*, 326.
30. Krishna, A.G.G.; Khatoon, S.; Shiela, P.M.; Sarmandal, C.V.; Indira, T.N.; Mishra, A. Effect or Reffining of Crude Rice Bran Oil on the Retention of Oryzanol in the Refined Oil. *J. Am. Oil Chem. Soc.* **2001**, *78*, 127–131. [CrossRef]

life

MDPI

Article

Argan Oil as a Rich Source of Linoleic Fatty Acid for Dietetic Structured Lipids Production

Tiago Simões [1,†], Jessica Ferreira [1,†], Marco F. L. Lemos [1], Ana Augusto [1,2,3], Rafael Félix [1], Susana F. J. Silva [1], Suzana Ferreira-Dias [4] and Carla Tecelão [1,4,*]

1 MARE-Marine and Environmental Sciences Centre, ESTM, Politécnico de Leiria, 2520-641 Peniche, Portugal; tiago.simoes@ipleiria.pt (T.S.); jessica_pinfer@yahoo.com (J.F.); marco.lemos@ipleiria.pt (M.F.L.L.); ana.l.augusto@ipleiria.pt (A.A.); rafael.felix@ipleiria.pt (R.F.); susana.j.silva@ipleiria.pt (S.F.J.S.)
2 CDRSP-Center for Rapid and Sustainable Product Development, Politécnico de Leiria, 2430-028 Marinha Grande, Portugal
3 Department of Food and Nutritional Sciences, University of Reading, Whiteknights, Reading RG6 6AP, UK
4 LEAF, Linking Landscape, Environment, Agriculture and Food, Instituto Superior de Agronomia, Universidade de Lisboa, 1349-017 Lisboa, Portugal; suzanafdias@isa.ulisboa.pt
* Correspondence: carla.tecelao@ipleiria.pt
† These authors contributed equally to this work.

Abstract: Argan oil is rich in long-chain unsaturated fatty acids (FA), mostly oleic and linoleic, and natural antioxidants. This study addresses the production of low-calorie structured lipids by acidolysis reaction, in a solvent-free system, between caprylic (C8:0; system I) or capric (C10:0; system II) acids and argan oil, used as triacylglycerol (TAG) source. Three commercial immobilized lipases were tested: Novozym® 435, Lipozyme® TL IM, and Lipozyme® RM IM. Higher incorporation degree (ID) was achieved when C10:0 was used as acyl donor, for all the lipases tested. Lipozyme® RM IM yielded the highest ID for both systems (28.9 ± 0.05 mol.% C10:0, and 11.4 ± 2.2 mol.% C8:0), being the only catalyst able to incorporate C8:0 under the reaction conditions for biocatalyst screening (molar ratio 2:1 FA/TAG and 55 °C). The optimal conditions for Lipozyme® RM IM in system II were found by response surface methodology (66 °C; molar ratio FA/TAG of 4:1), enabling to reach an ID of 40.9 mol.% of C10:0. Operational stability of Lipozyme® RM IM in system II was also evaluated under optimal conditions, after eight consecutive 24 h-batches, with biocatalyst rehydration between cycles. The biocatalyst presented a half-life time of 103 h.

Keywords: *Argania spinosa* oil; capric acid; caprylic acid; commercial immobilized lipases; low-calorie structured lipids

Citation: Simões, T.; Ferreira, J.; Lemos, M.F.L.; Augusto, A.; Félix, R.; Silva, S.F.J.; Ferreira-Dias, S.; Tecelão, C. Argan Oil as a Rich Source of Linoleic Fatty Acid for Dietetic Structured Lipids Production. *Life* **2021**, *11*, 1114. https://doi.org/10.3390/life11111114

Academic Editor: Othmane Merah

Received: 22 September 2021
Accepted: 14 October 2021
Published: 20 October 2021

1. Introduction

Lipids are important diet constituents due to their high energy value and essential fatty acid content, being involved in crucial mechanisms such as cell signalling and hormonal regulation [1]. Triacylglycerols (TAG), the main lipid components of dietary fat, consist of a glycerol backbone linked to three fatty acids. Depending on their structure and composition, they present distinct characteristics such as melting behaviour, digestion, absorption, and metabolic properties [2,3].

Structured lipids (SL), also known as "taylor-made fats", usually consist of TAG with a chemically or enzymatically modified structure, obtained by changing the original fatty acid (FA) profile and/or their original position in the glycerol structure [4–6]. Enzymatic synthesis of structured lipids presents several advantages in comparison with the chemical route namely milder reaction conditions, enantio- and regioselectivity, less toxicity and oxidation, as well as easier product recovery, among others [3,7,8]. Despite not being characteristic for all structured lipids, some of these modified molecules have demonstrated their nutritional and functional properties in treating and preventing particular human diseases [1,9]. In recent years, research focused on low-calorie (dietetic) fats has increased, to meet consumers' demand for a healthier lifestyle and to reduce obesity. Structured triacylglycerols presenting low caloric value comprise TAG containing short or medium-chain FA (C_8–C_{12}), preferably esterified at external *sn*-1,3 positions, and long-chain FA (L, $\geq C_{14}$) occupying the internal position of the glycerol backbone. These structured lipids, named MLM, provide health and nutrition benefits since they may control various metabolic disorders, such as obesity and fat malabsorption [2,10,11]. The average caloric value reported for these fats is 5 kcal g^{-1}, while conventional counterparts present 9 kcal g^{-1} [6,12–14]. During digestion, the *sn*-1,3-regioselective pancreatic lipase releases the medium-chain FA from TAG, which are preferentially transported to the liver where they are promptly metabolized via mitochondrial β-oxidation, providing an energy source without accumulation in the adipose tissue [15]. The effective incorporation of the resulting *sn*-2-monoacylglycerols (2-MAG) into chylomicrometers and its absorption through the lymphatic system is observed [7].

MLM structured lipids are currently synthesized by acidolysis reaction between an oil rich in long-chain FA (e.g., olive, olive pomace, linseed, spent coffee grains, grapeseed, avocado, or microbial oils) and medium-chain FA, namely octanoic (C8:0; common name—caprylic acid) or decanoic (C10:0; common name—capric acid) acids, catalyzed by *sn*-1,3-regioselective lipases [10,12,13,16–24].

Lipases (E.C. 3.1.1.3., triacylglycerol acylhydrolases) are versatile enzymes able to catalyze the hydrolysis of ester bonds in the aqueous phase and promote the reverse reaction of esterification, as well as interesterification and transesterification reactions, in medium with low water activity [25].

Several immobilized lipases have been tested and used for MLM production, presenting high yields and operational stability in solvent-free systems [4,13]. Implementing cost-effective reaction systems for MLM production may be attained by the use of less expensive biocatalysts, with high activity and operational stability, as well as oils with valuable properties to be used as a TAG source [26–28]. In this sense, virgin argan oil may be a feasible substrate due to its anti-inflammatory, antioxidant, and cardiovascular-protective properties. Argan oil (AO) contains natural antioxidants, namely tocopherols and other sterols and phenolic compounds [29–32]. These compounds provide higher oxidative stability to the argan oil when compared to other oils [33]. They may also be beneficial in the reaction medium, by preventing or delaying lipid oxidation. TAG of argan oil mainly contain oleic (O, C18:1 n-9), linoleic (L, C18:2 n-6) and palmitic (P, C16:0) fatty acids. Therefore, the main TAG are OOL, POL, OLL and OOO [34]. The TAG structure, containing mainly oleic or linoleic acid at position *sn*-2, and the antioxidant composition turn this oil very appealing to be used in structured lipids synthesis, namely, for MLM production.

This study aimed at producing MLM by lipase-catalyzed acidolysis between medium-chain FA (caprylic or capric acids) and TAG from argan oil, in the absence of organic solvents. Three commercial biocatalysts were tested for this purpose. The lipase showing the highest activity was selected for modelling and optimization of reactions conditions (molar ratio (FA/TAG) and temperature).

2. Materials and Methods

2.1. Biocatalysts, Substrates, and Chemicals

The immobilized thermostable lipases from *Candida antarctica* lipase b (Novozym® 435), *Rhizomucor miehei* (Lipozyme® RM IM) and *Thermomyces lanuginosus* (Lipozyme® TL IM), were kindly provided by Novozymes, A/S, Bagsvaerd, Denmark. Edible virgin argan oil (Emile Noël, France) was acquired from a biological food store. Caprylic (>98% purity; MW = 144.21 g mol^{-1}) and capric (>98% purity; MW = 172.26 g mol^{-1}) fatty acids, and methyl myristate standard (99%) were acquired from Fluka. Silica-gel 60 (0.25 mm width, 20 × 20 cm) thin layer chromatography (TLC) plates, acetyl chloride, methanol and *n*-heptane were acquired from Merck, Germany, and 2′,7′-dichlorofluorescein was from

Sigma-Aldrich. Fatty acid methyl ester mixes (PUFA No 1 from Marine source and PUFA No 3 from Menhaden oil) were acquired from Supelco (Bellefonte, PA, USA).

2.2. Biocatalyst Screening

Acidolysis reactions were carried out batchwise, for all the tested biocatalyst, using cylindrical jacketed glass reactors (20 mL) operating at 55 °C, under magnetic stirring of 300 rpm. Rubber caps were used to seal the reactors to minimize oxidation in the reaction system. The reaction medium comprised 1.95 g of AO, as the TAG source, and 1.35 g of C8:0 (system I) or C10:0 (system II) fatty acids, resulting in a molar ratio FA/AO of approximately 2:1, calculated based on the molecular weight of trilinolein. This is the stoichiometric ratio required for the esterification of FA at the external positions of the glycerol backbone by *sn*-1,3 selective lipases. A 5% (*w*/*w*) lipase load was used, with respect to the amount of oil. After 24 h reaction, the enzyme was recovered from the medium by filtration (Whatman™ Grade 1 filter paper) at room temperature. All assays were carried out in duplicate. The filtrate was stored at −20 °C for subsequent analyses.

2.3. Time-Course Experiments

The biocatalyst presenting the highest acidolysis activity was used in a time-course reaction to evaluate when a quasi-equilibrium was reached. Thus, an acidolysis reaction between AO and C10:0 (system I), catalysed by Lipozyme RM IM, was performed for 48 h, with withdraw samples along the reaction. The assay was carried out under the same conditions as described in Section 2.2, corrected to a total reaction volume of 7 mL. Duplicate aliquots of 200 μL were collected, under stirring, in every sampling for TAG composition analysis.

2.4. Modelling Acidolysis and Optimization of Reaction Conditions

For acidolysis modelling and reaction conditions optimization, as a function of the molar ratio FA/AO and temperature, a set of experiments was performed under the conditions dictated by a central composite rotatable design (CCRD) [35]. In this experimental design, each variable (molar ratio and temperature) was tested at five different levels to determine the effects of each one on MLM production. Experiments were conducted using Lipozyme RM IM lipase for 24 h, following the methodology described in Section 2.2. The lipase load was kept constant at 5% (*w*/*w*). The CCRD consisted of 11 assays, as a function of temperature (T: 44–66 °C) and FA/AO molar ratio (MR: 1.2:1–6.8:1). Results were computed with Statistica software, version 10 (TIBCO, Tulsa, CA, USA). The influence of the factor's temperature and substrate molar ratio on the response (C10:0 incorporation) was calculated, considering the linear and quadratic effects as well as their linear interaction. Analysis of variance evaluated their significance, being considered statistically significant at level 0.05 ($p < 0.05$). A three-dimensional surface, characterized by a second-order polynomial equation, was adjusted to the C10:0 incorporation values obtained in CCRD experiments [35]. The statistical principle of least squares was used to estimate the first and second-order coefficients of this equation from the experimental data. The determination coefficients (R^2) and adjusted R^2 ($Radj^2$) were used to evaluate the goodness of fit of the model. To validate the model, the acidolysis of AO with C10:0, catalysed by RM IM lipase, was carried out for 24 h under the optimized conditions dictated by the model.

2.5. Batch Operational Stability Assays

Operational stability tests comprised eight successive 24 h batch reactions, at the predicted optimal conditions: 66 °C and a molar ratio of 4:1 (FA/AO). At the end of each batch, the lipase was removed by paper filtration, at room temperature, hydrated with 50 mL of a solution of phosphate buffer (pH 7; 0.1 M) and dried by filtration under vacuum to remove the excess buffer solution. Then, the recovered lipase was added to a fresh medium and reused in a new batch under the same reaction conditions. The catalytic activity of the lipase was considered as the observed molar incorporation degree of capric acid in argan oil TAG, at the end of each batch. The activity of the biocatalyst exhibited at the end of the first batch is considered 100% activity. After each reuse, the residual activity was expressed as a percentage of the activity in batch one.

The fit of the inactivation models was performed using the "Solver" add-in from Excel for Windows, by minimizing the residual sum-of-squares between the experimental results and those estimated by the models.

2.6. *Reaction Products Analysis*

All reaction products (FFA, partial acylglycerols and TAG) were separated by silica gel thin-layer chromatography (TLC). The TAG band was recovered, methylated and analysed by gas chromatography (GC), as fatty acid methyl esters (FAME), according to the methodology described by [36]. A gas chromatograph Finnigan TRACE GC Ultra (Thermo Electron Corporation) with a capillary column (60 m × 0.25 mm ID × 0.25 µm film) Thermo TR-FAME, and an AS 3000 autosampler from Thermo Electron Corporation were used for FAME analysis. The flame ionization detector (FID) was set at 280 °C and supplied with air and hydrogen at 350 and 35 mL min^{-1}, respectively, The injector (in splitless mode) was set at 250 °C. The temperature program of the chromatographic column was the following: 100 °C for 1 min, increase to 160 °C at 10 °C min^{-1}, held for 10 min, increase to 235 °C at 4 °C min^{-1} and kept for 10 min. The carrier gas Helium was supplied at a flow rate of 1.2 mL min^{-1}. Fatty acid methyl ester mixes (PUFA No1 and PUFA No 3, from Supelco) were used as external standard and methyl myristate (C14:0) as internal standard.

The incorporation degree (*ID*) of C8:0 or C10:0 in TAG of argan oil, was calculated as follows, Equation (1) [20]:

$$ID\ (\%) = \left(\frac{\text{MFA}}{\text{MT}}\right) \times 100 \tag{1}$$

where MFA represents the C8:0 or C10:0 moles in the TAG and MT is the total amount of fatty acids (moles) in the TAG.

3. Results and Discussion

3.1. *Fatty Acid Composition of Argan Oil*

The fatty acid composition of argan oil used in this study is presented in Table 1, as well as the reference values following the Moroccan norm (N.M. 08.5.090), established for edible virgin argan oil.

The fatty acid profile of argan oil used in the present study fulfils the requirements established by the Moroccan norm. Oleic acid is the major fatty acid (41.1 ± 2.90), followed by linoleic (35.9 ± 1.40), palmitic (14.31 ± 0.63) and stearic (6.18 ± 0.19) acids.

3.2. *Biocatalyst Screening*

The commercial immobilized lipases Novozym® 435, Lipozyme® TL IM and Lipozyme® RM IM were tested as catalysts for the acidolysis reaction between AO and C8:0 (system I) or C10:0 (system II) fatty acids, to produce MLM-type structured lipids. The tested enzymes were selected due to their (i) *sn*-1,3 regioselectivity, (ii) high activity in both aqueous and non-conventional media (aqueous/organic), (iii) ability to operate either in batch as in continuous reactors, (iv) stability at a wide range of pH and temperature, (v) operational stability, dependent on the reaction conditions and (vi) application at large-scale production. [4,9]. However, some studies indicate that Novozym® 435 may present non-specific regioselectivity, depending on the reaction medium composition or immobilization supports [36–39].

Table 1. Comparative analysis of fatty acid profile of argan oil used in the present study and the official values established in the Moroccan norm (N.M. 08.5.090). Samplings were performed in duplicate and results are presented in percentage, with standard deviation (S.D.). C represents the number of carbon in the fatty acid and D is the number of double bonds in the carbon chain; SFA: saturated fatty acids; MUFA: monounsaturated fatty acids; PUFA: polyunsaturated fatty acids.

Fatty Acid	Lipid Number (C:D)	Percentage (% ± S.D.)	Normative Composition (Charrouf and Guillaume 2014)
Myristic	C 14:0	0.20 ± 0.01	≤0.2
Palmitic	C 16:0	14.31 ± 0.63	11.5–15.0
Palmitoleic	C 16:1 n9	0.14 ± 0.01	≤0.2
Stearic	C 18:0	6.18 ± 0.19	4.3–7.2
Oleic	C 18:1, n9	41.1 ± 2.90	43.0–49.1
Linoleic	C 18:2 n6	35.9 ± 1.40	29.3–36.0
α-Linolenic	C 18:3 n3	0.125 ± 0.01	≤0.3
Arachidic	C 20:0	0.31 ± 002	≤0.5
Gadoleic	C 20:1 n9	0.34 ± 0.01	≤0.5
Behenic	C 22:0	0.1 ± 0.001	≤0.2
∑ SFA	-	21.2 ± 0.87	-
∑ MUFA	-	42.4 ± 2.34	-
∑ PUFA	-	36.27 ± 1.44	-

The structured lipids (MLM type) were synthesized in a solvent-free medium which has several advantages, namely: enables its direct incorporation in food matrices without the need for solvent removal and SL purification, reduces the production costs, besides being a more environmentally friendly process. Additionally, the presence of an organic solvent will dilute the reaction medium leading to lower yields in MLM, even when high incorporation levels of medium-chain fatty acids are observed.

Results from the initial screening, to evaluate the most suitable biocatalyst, are presented in Figure 1. In this study, Lipozyme® RM IM achieved the highest incorporation of C10:0 (28.9 ± 0.05 mol.%), being the only biocatalyst able to integrate C8:0 in argan oil, under the screening conditions (11.4 ± 2.2 mol.%).

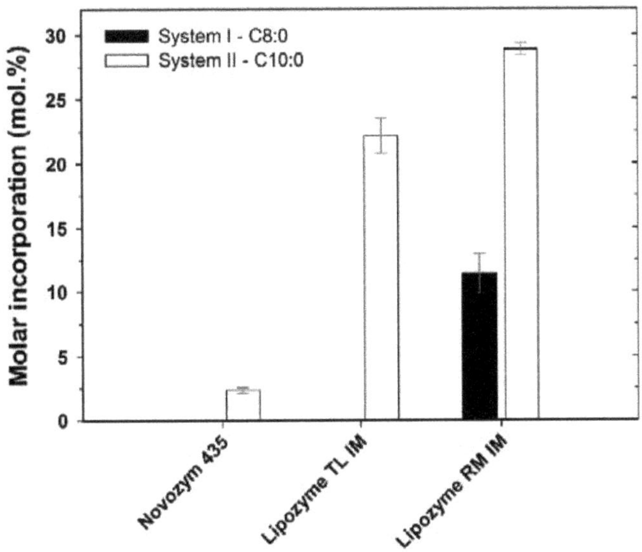

Figure 1. Molar incorporation of caprylic acid (System I. C8:0) or capric acid (System II C10:0) in the argan oil, using different commercial biocatalysts, at 55 °C after 24 h acidolysis in a solvent-free medium.

Nunes et al. [23] used a similar approach for evaluating both C8:0 or C10:0 incorporations into virgin olive oil, catalysed by the same commercial immobilized lipases used in the present study. Two different reaction systems were tested (solvent vs. solvent-free medium) and higher incorporation degrees of C10:0 (ranging from 27.1 to 30.4 mol.%) were also observed in a solvent-free medium in comparison with C8:0 (19.9 to 25.7 mol.%), for all the biocatalysts tested. Low incorporations of caprylic acid into soybean oil (SBO), using Lipozyme TL IM as biocatalyst in the absence of organic solvent, were also reported by Turan et al. [40]. The highest yield (7.84%) was obtained at 40 °C, for an enzyme load of 10 wt.%, reaction time of 6 h and a molar ratio of 1:0.7 (SBO:C8:0). Kim et al. [41] reported higher incorporation of caprylic acid in perilla oil, by acidolysis catalysed by Lipozyme TL IM (30.6 mol.%) and by Lipozyme RM IM (34.2 mol.%), at 55 °C, using a molar ratio of 1:6, in solvent-free media. In the acidolysis reaction between C8:0 and avocado oil (rich in C18:1), catalysed by Lipozyme RM IM and Lipozyme TL IM, the highest incorporation value (29.2 mol.%) was attained for Lipozyme TL IM, operating at 30 °C along 24 h [12].

The type of oil, the biocatalyst and the operating conditions used are shown to greatly affect the success of acidolysis to synthesize low-calorie TAG. Considering the highest molar incorporation achieved with argan oil, the reaction system containing capric acid as acyl donor and using Lipozyme® RM IM as catalyst was selected for subsequent studies.

3.3. Time-Course Reaction

The evaluation of the enzymatic reaction kinetics is very important to select the time required to attain a quasi-equilibrium situation. Therefore, the incorporation of C10:0 in TAG from AO was monitored along time, under the same conditions followed for biocatalyst and system screening. The incorporation degree along 48 h acidolysis reaction is presented in Figure 2.

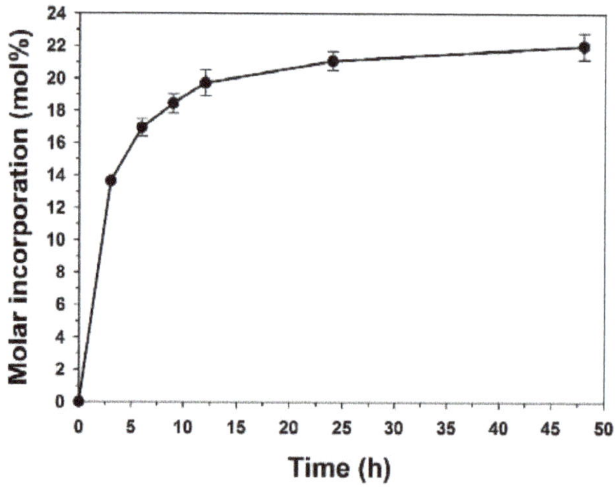

Figure 2. Incorporation (performed in duplicate) of C10:0 in the argan oil (molar ratio C10:0/AO of 2:1), catalyzed by 5% of Lipozyme® RM IM (*w:w* of AO), in the absence of organic solvent, at 55 °C, during 48 h.

A quasi-equilibrium was observed between 24 and 48 h of reaction. After 24 h, C10:0 molar incorporation was averaged in 21.2 mol.% and 22.0 mol.% after 48 h acidolysis. Similar results of capric acid incorporation by acidolysis in solvent-free systems, were previously reported, with virgin olive oil (27.1 ± 2.3 mol.%) and Lipozyme RM IM after 24 h [23], with *Cucurbita maxima* pumpkin seed oil (29.9 ± 0.7 mol.%) and Lipozyme TL IM, at 45 °C, after 31 h acidolysis [42].

3.4. Modelling Acidolysis and Optimization of Reaction Conditions

In order to optimize the operational conditions that maximize capric acid incorporation in TAG of argan oil, the influence of the factors substrate molar ratio and temperature were evaluated through a set of acidolysis reactions following a CCRD. The experimental conditions and the obtained results are described in Table 2.

Table 2. Experimental matrix, as a function of molar ratio (MR, C10:0/AO) and temperature (T; °C), and the results obtained for molar incorporation of C10:0 in TAG from argan oil, after 24 h acidolysis catalyzed by Lipozyme® RM IM.

Experiments	Coded Matrix		Decoded Matrix		Incorporation (mol.%)
	Temperature (°C)	Molar Ratio (C10:0/AO)	Temperature (°C)	Molar Ratio (C10:0/AO)	
1	−1	−1	44	2:1	27.7
2	−1	1	44	6:1	35.4
3	1	−1	62	2:1	36.9
4	1	1	62	6:1	39.7
5	−1.414	0	40	4:1	32.1
6	1.414	0	66	4:1	40.9
7	0	−1.414	53	1.2:1	19.0
8	0	1.414	53	6.8:1	37.8
9	0	0	53	4:1	34.7
10	0	0	53	4:1	37.2
11	0	0	53	4:1	32.5

The highest incorporation degree (40.9 mol.%) was obtained at the highest temperature tested (66 °C) and a molar ratio to 4:1 C10:0/AO. Conversely, the lowest incorporation was observed for the lowest MR tested (1.2:1) which is below the stoichiometric value of 2:1. A negative or a positive linear effect of a factor (MR or T) on capric acid incorporation means a reduction or increase in the response, with the increase in the factor value, respectively. A positive or negative quadratic effect indicates that the response is described by a concave or convex response surface, respectively. The influence of MR and T on the acidolysis reaction was studied by evaluating both linear and quadratic effects of each factor, as well as their linear interaction on C10:0 incorporation in AO (Table 3).

Table 3. Linear and quadratic effects of molar ratio (MR), temperature (T)and linear interaction (T X MR) in the incorporation of C10:0 in argan oil. with the corresponding *p* values.

Factor	Effect	*p*-Value
T (linear)	6.41	0.043
T (quadratic)	2.81	0.357
MR (linear)	9.30	0.012
MR (quadratic)	−5.19	0.135
T X MR (linear)	−2.45	0.503

The incorporation of C10:0 was influenced by both molar ratio and temperature. A positive linear effect was observed for temperature, being statistically significant ($p = 0.043$). This suggests that increasing temperature will result in increased incorporations. In fact, a temperature increase promotes the conversion of substrates into products by improving substrates solubility, decreasing reaction medium viscosity and favouring mass transfer [43,44]. The linear effect observed for MR was also positive and significant, suggesting the same outcome as for temperature. The quadratic effect of MR is negative with a *p*-value of 0.135, which is above the limit of 0.05. However, this effect must be considered because its removal would lead to a lack of fit of the polynomial model adjusted to the experimental data, which would be confirmed by a considerable decrease in both R^2 and R^2adj of that model. In fact, is preferred to accept factors presenting effects with *p*-values higher than 0.05 rather than to ignore some important factors [45]. The negative quadratic effect of

MR indicates a convex response surface describing the incorporation of C10:0 in AO, as a function of MR. The interaction effect of T X MR is not significant ($p = 0.503$) and, therefore, can be ignored in the response surface model. In this study, C10:0 incorporation in the TAG followed a convex surface as a function of the molar ratio (indicated by the negative MR quadratic effect).

The incorporation of C10:0 in TAG of AO, *ID*, as a function of both T and MR, catalyzed by Lipozyme® RM IM, may be represented by a convex surface (Figure 3), described by the following second-order polynomial Equation (2):

$$D\ (\text{mol.\%}) = 44.92 - 1.48\text{T} + 0.017\ \text{T}^2 + 7.52\ \text{MR} - 0.649\text{MR}^2 \tag{2}$$

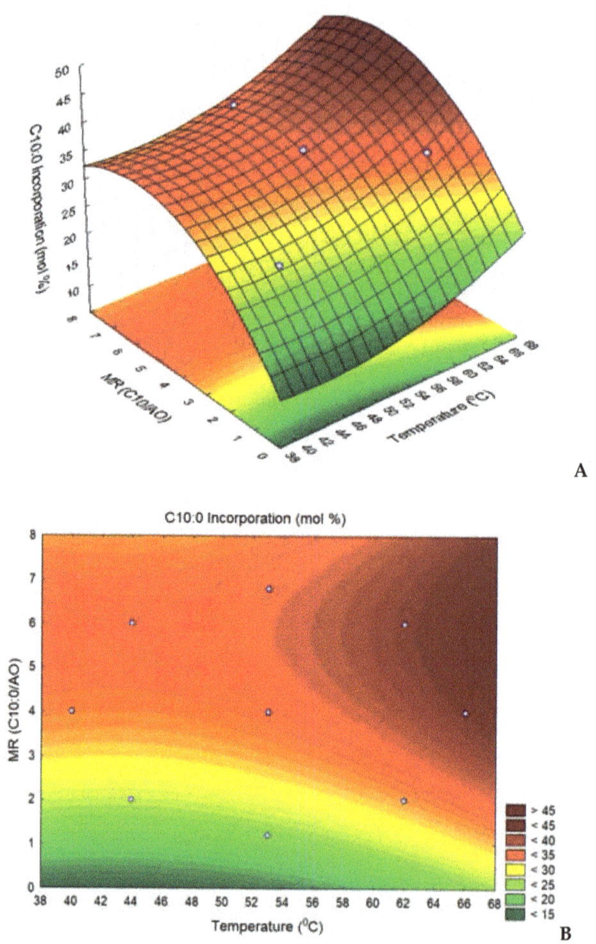

Figure 3. Three-dimensional response surface (**A**) describing the incorporation of capric acid (mol.%) in the argan oil, and (**B**) respective projection, as a function of reaction temperature and the molar ratio, MR, capric acid/argan oil (C10:0/AO).

The values of the coefficient of determination (R^2) and the adjusted coefficient of determination (R^2_{adj}) were 0.835 and 0.725, respectively. This shows a good fit of the model to the experimental data, proving them to be predictable by this model.

Figure 3 shows that the incorporation of C10:0 in argan oil increases with T but presents an optimal MR outside the experimental domain chosen in this study. Yet, it is possible to identify the region corresponding to the best results (T = 64–66 °C; MR = 4–5:1). In the acidolysis reaction between C8:0 or C10:0 and grapeseed oil, in the solvent-free system, using the same immobilized commercial lipases, Bassan et al. [19] reported an optimal temperature of 69 °C in the reaction system, also found by Response Surface Methodology, which was very close to the optimal temperature observed in this study (Figure 3). Additionally, the best results were reported for C10:0 as medium-chain fatty acid and Lipozyme RM IM lipase as a catalyst. High temperatures may compromise the activity and the operational stability of the enzyme (discussed in Section 3.5). The use of high MR in the reaction medium often presents no significant, or even a negative effect on SL synthesis, which may be attributed to an inhibition of the enzyme by the substrate and/or to a lipase deactivation due to the high levels of FFA in the reaction medium. This is sustained by several reports in the literature where increasing concentrations of FFA resulted in lower incorporations in the TAG backbone. Concerning the synthesis of human milk fat substitutes, lower incorporations of *n*3-PUFAs in the TAG structure of lard [26] and tripalmitin [46] were observed by increasing their concentration in the reaction medium, using the immobilized lipases of heterologous *Rhizopus oryzae* and *Carica papaya* as biocatalyst, respectively. Aiming at the production of low caloric TAG, [40] observed higher incorporations of caprylic acid into soybean oil at lower molar ratios, catalyzed by Lipozyme TL IM, being the highest value achieved at an MR of 1:0.7 (SBO:CA). Additionally, Nunes et al. [16] reported similar observations for the acidolysis of virgin olive oil with C8:0 or C10:0 fatty acids, catalysed by *Rhizopus oryzae*, with optimal molar ratios of 2.8:1 for C8:0/TAG and 3:1 for C10:0/TAG, respectively. These observations are aligned with the present results, reporting that increasing substrate molar ratio excessively may translate into lower yields. Such results are important to consider from an industrial point of view, since the use of large amounts of substrates will increase operational costs related to product and unconverted substrate recovery.

3.5. Operational Stability

The low operational stability is identified as one of the main drawbacks to the scale-up of enzyme-catalyzed processes to industrial applications. In this study, the operational stability assays were carried out at the best conditions predicted by the CCRD model, i.e., at 66 °C and a molar ratio of 4:1 (C10:0/AO). Eight consecutive 24 h batches were performed with Lipozyme® RM IM, rehydrated between every two consecutive batches, accounting for a total of 192 operating hours. The results for residual activity after consecutive batches are presented in Figure 4.

Hydration favours the catalytic activity and prevents enzyme denaturation during dehydration in a non-aqueous reaction system [47]. Rehydration was carried out since it previously showed positive effects on the operational stability of r-ROL immobilized in Lewatit VP OC 1600, used as catalyst in the acidolysis of virgin olive oil with capric acid [24]. Conversely, negative effects of rehydration between batches on lipase activity, which was related to the enzyme itself and the carrier used for immobilization, have been reported [48].

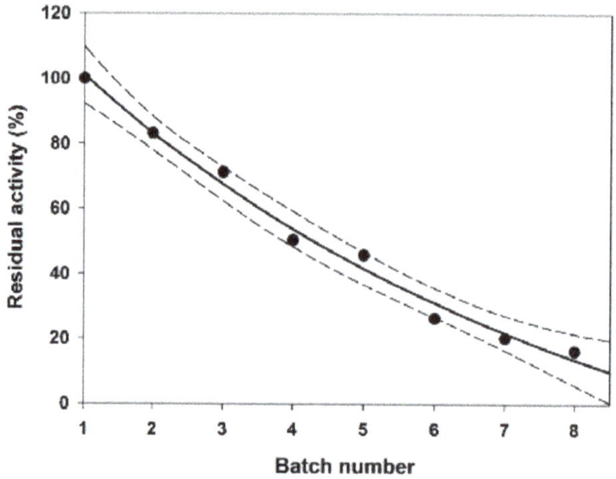

Figure 4. Lipozyme® RM IM residual activity (%) at the end of each 24 h reuse, during acidolysis reaction between C10:0 and argan oil, at 66 °C and molar ratio of 4:1 (FA:TAG). Confidence intervals (95%) presented in dash lines.

A decrease in the biocatalyst residual activity was observed between each consecutive batch. After 96 h operation, Lipozyme RM IM retained 53% of its original activity. Deactivation results were best fitted to a first-order deactivation kinetics model, given by Equation (3).

$$A_n = Ae^{-k_d n} \tag{3}$$

where A_n is the residual activity (%) of the biocatalyst at the end of batch n, A is a constant that represents the enzyme initial activity before deactivation, and k_d is the deactivation rate constant, expressed in n^{-1}. The following Equation (4) represents the fit of this model to the experimental data:

$$A_n = 165.74e^{-0.131n} \tag{4}$$

Data showed a good fitting to the model (R^2adj = 0.989), with a half-life time ($t_{1/2}$) of 103 h, corresponding to approximately 4.3 days. These results can be compared to some reports in the literature, where similar operating systems and the same biocatalysts were tested. Lipozyme RM IM presented higher operational stability in (i) the continuous interesterification between palm stearin (45% *w*/*w*), palm kernel oil (30% *w*/*w*) and olive oil (25% *w*/*w*), using a packed-bed bioreactor at 65 °C [49], (ii) the interesterification of tripalmitin with oleic acid or omega-3 polyunsaturated fatty acids (omega-3 PUFA) as acyl donors, aiming at the production of human milk fat substitutes [36], and (iii) the synthesis of MLM structured lipids by acidolysis of virgin olive oil with caprylic or capric fatty acids [23].

Lipozyme® RM IM is immobilized in a macroporous anion exchange resin, which confers thermal stability to the biocatalyst. However, extended operating periods at high temperatures may cause lipase inactivation and its release from the support, or the opposite situation, release from the support and consequent inactivation [50].

Although it is a thermostable lipase, when contrasting the present results with the literature, a credible better performance of Lipozyme RM IM at milder temperatures for more prolonged activity periods may be expected. This was previously reported for similar reaction systems [12,23]. This may also be related to the inactivation of the biocatalyst by lipid oxidation compounds formed at high temperatures, more than biocatalyst preferences for specific length FA. A cost-effective analysis of operating conditions (stability vs. temperature) would be necessary to predict the optimal reaction conditions for MLM production using Lipozyme RM IM as catalyst. It has also been reported that the purity of the substrates influences the stability of biocatalysts. Minor natural compounds in the reaction medium, such as phospholipids, emulsifiers, chlorophyll, carotenoids, or antioxidants, remarkably affect the stability of lipases [51]. Virgin argan oil contains about 1% of these molecules. Natural molecules such as tocopherols and polyphenols were not evaluated in the present study, but initially, their antioxidant activity would be expected in the reaction medium due to their biochemical nature, interfering with lipid autoxidation by a prompt donation of hydrogen atoms to lipid radicals [4]. However, the relatively high optimal temperature (66 °C), along with the reported negative effects for these compounds may have adversely affected the biocatalyst stability in this study. An additional characterization of these molecules in the oil, along with its oxidative stability, would be a compelling prospective goal.

4. Conclusions

This study showed that argan oil may be a feasible substrate for MLM-type structured lipid production. In fact, using this oil as TAG source in the acidolysis reaction with capric acid (C10:0), it was possible to achieve an incorporation degree of 40.9 mol.% at the optimal conditions dictated by the response surface methodology (66 °C and a molar ratio 4:1 C10:TAG), after 24 h. This was achieved by using Lipozyme RM IM, which showed to be the best biocatalyst comparatively to other commercial lipases tested in this reaction system. The biocatalyst presented a half-life time ($t_{1/2}$) of 103 h, after eight consecutive 24 h batches, at the same reaction conditions. Operating conditions such as MR and temperature are important factors to consider in similar operating systems to further implementation at a larger scale.

Author Contributions: Conceptualization, M.F.L.L., S.F.J.S., S.F.-D. and C.T.; methodology, T.S., J.F., M.F.L.L., A.A., R.F., S.F.J.S., S.F.-D. and C.T.; laboratory work.; T.S. and J.F.; resources, M.F.L.L., S.F.-D. and C.T.; data curation, T.S., J.F., S.F.-D. and C.T.; writing—original draft preparation, T.S. and J.F.; writing—review and editing, M.F.L.L., A.A., R.F., S.F.J.S., S.F.-D. and C.T.; supervision, C.T.; project administration, M.F.L.L., S.F.J.S. and C.T.; funding acquisition, M.F.L.L., S.F.J.S. and C.T. All authors have read and agreed to the published version of the manuscript.

Funding: This study had the support of Fundação para a Ciência e a Tecnologia (FCT), Portugal, to the research units of MARE (UID/MAR/04292/2020) and LEAF-Linking Landscape, Environment, Agriculture and Food Research Centre (UIDB/04129/2020), and of the grants awarded to Tiago Simões (SFRH/BD/98266/2013), Ana Augusto (SFRH/BD/131465/2017) and Rafael Félix (SFRH/ BD/139763/2018). The authors also acknowledge the project ValorMar (24517, 10/SI/2016-Mobilizing R&TD Programs, Portugal 2020), co-funded by the European Regional Development Fund (ERDF) and by the European Social Fund (ESF). The work was also partially funded by the Integrated Programme of SR&TD "SmartBioR" (reference Centro-01-0145-FEDER-000018) co-funded by Centro 2020 program, Portugal2020, European Union, through the European Regional Development Fund.

Conflicts of Interest: The authors declare no conflict of interest.

References

1. Ibarguren, M.; López, D.J.; Escribá, P.V. The effect of natural and synthetic fatty acids on membrane structure, microdomain organization, cellular functions and human health. *Biochim. Biophys. Acta Biomembr.* **2014**, *1838*, 1518–1528. [CrossRef] [PubMed]
2. Timm-Heinrich, M.; Xu, X.; Nielsen, N.S.; Jacobsen, C. Oxidative stability of milk drinks containing structured lipids produced from sunflower oil and caprylic acid. *Eur. J. Lipid Sci. Technol.* **2003**, *105*, 459–470. [CrossRef]
3. Kadhum, A.A.H.; Shamma, M.N. Edible lipids modification processes: A review. *Crit. Rev. Food Sci. Nutr.* **2017**, *57*, 48–58. [CrossRef]
4. Ferreira-Dias, S.; Sandoval, G.; Plou, F.; Valero, F. The potential use of lipases in the production of fatty acid derivatives for the food and nutraceutical industries. *Electron. J. Biotechnol.* **2013**, *16*, 1–32. [CrossRef]
5. Ferreira-Dias, S.; Osório, N.M.; Rodrigues, J.; Tecelão, C. Structured lipids for foods. In *Encyclopedia of Food Chemistry*; Varelis, P., Melton, L., Shahidi, F., Eds.; Elsevier: Amsterdam, The Netherlands, 2018; pp. 357–369. ISBN 9780128140260.
6. Kim, B.H.; Akoh, C.C. Recent research trends on the enzymatic synthesis of structured lipids. *J. Food Sci.* **2015**, *80*, C1713–C1724. [CrossRef] [PubMed]
7. He, Y.; Li, J.; Guo, Z.; Chen, B. Synthesis of novel medium-long-medium type structured lipids from microalgae oil via two-step enzymatic reactions. *Process Biochem.* **2018**, *68*, 108–116. [CrossRef]

8. Rios, N.S.; Pinheiro, B.B.; Pinheiro, M.P.; Bezerra, R.M.; dos Santos, J.C.S.; Barros Gonçalves, L.R. Biotechnological potential of lipases from Pseudomonas: Sources, properties and applications. *Process Biochem.* **2018**, *75*, 99–120. [CrossRef]
9. Lopes, P.A.; Pestana, J.M.; Coelho, D.; Madeira, M.S.; Alfaia, C.M.; Prates, J.A.M. From natural triacylglycerols to novel structured lipids containing n-3 long-chain polyunsaturated fatty acids. In *The Molecular Nutrition of Fats*; Patel, V., Ed.; Academic Press: Cambridge, MA, USA, 2018; pp. 225–235. [CrossRef]
10. Paul Martínez-Galán, J.; María Ontibón-Echeverri, C.; Campos Costa, M.; Batista-Duharte, A.; Guerso Batista, V.; Mesa, V.; Monti, R.; Veloso De Paula, A.; Martins Baviera, A. Enzymatic synthesis of capric acid-rich structured lipids and their effects on mice with high-fat diet-induced obesity. *Food Res. Int.* **2021**, *148*, 963–9969. [CrossRef]
11. Lu, J.; Jin, Q.; Wang, X.; Wang, X. Preparation of medium and long chain triacylglycerols by lipase-catalyzed interesterification in a solvent-free system. *Process Biochem.* **2017**, *54*, 89–95. [CrossRef]
12. Caballero, E.; Soto, C.; Olivares, A.; Altamirano, C. Potential use of avocado oil on structured lipids MLM-type production catalysed by commercial immobilised lipases. *PLoS ONE* **2014**, *9*, e107749. [CrossRef]
13. Costa, C.M.; Osório, N.M.; Canet, A.; Rivera, I.; Sandoval, G.; Valero, F.; Ferreira-Dias, S. Production of MLM type structured lipids from grapeseed oil catalyzed by non-commercial lipases. *Eur. J. Lipid Sci. Technol.* **2018**, *120*, 1–8. [CrossRef]
14. Akoh, C.C.; Pande, G. Structured Lipids and Health. In *Bailey's Industrial Oil and Fat Products*; Shahidi, F., Ed.; Wiley: Hoboken, NJ, USA, 2020; pp. 329–346. [CrossRef]
15. Schönfeld, P.; Wojtczak, L. Short- and medium-chain fatty acids in energy metabolism: The cellular perspective. *J. Lipid Res.* **2016**, *57*, 943–954. [CrossRef] [PubMed]
16. Nunes, P.A.; Pires-Cabral, P.; Guillén, M.; Valero, F.; Ferreira-Dias, S. Batch operational stability of immobilized heterologous Rhizopus oryzae lipase during acidolysis of virgin olive oil with medium-chain fatty acids. *Biochem. Eng. J.* **2012**, *67*, 265–268. [CrossRef]
17. Mota, D.A.; Santos, J.C.B.; Faria, D.; Lima, Á.S.; Krause, L.C.; Soares, C.M.F.; Ferreira-Dias, S. Synthesis of dietetic structured lipids from spent coffee grounds crude oil catalyzed by commercial immobilized lipases and immobilized Rhizopus oryzae lipase on biochar and hybrid support. *Processes* **2020**, *8*, 1542. [CrossRef]
18. Abed, S.M.; Wei, W.; Ali, A.H.; Korma, S.A.; Mousa, A.H.; Hassan, H.M.; Jin, Q.; Wang, X. Synthesis of structured lipids enriched with medium-chain fatty acids via solvent-free acidolysis of microbial oil catalyzed by Rhizomucor miehei lipase. *LWT* **2018**, *93*, 306–315. [CrossRef]
19. Bassan, N.; Rodrigues, R.H.; Monti, R.; Tecelão, C.; Ferreira-Dias, S.; Paula, A.V. Enzymatic modification of grapeseed (Vitis vinifera L.) oil aiming to obtain dietary triacylglycerols in a batch reactor. *LWT* **2019**, *99*, 600–606. [CrossRef]
20. Casas-Godoy, L.; Marty, A.; Sandoval, G.; Ferreira-Dias, S. Optimization of medium chain length fatty acid incorporation into olive oil catalyzed by immobilized Lip2 from Yarrowia lipolytica. *Biochem. Eng. J.* **2013**, *77*, 20–27. [CrossRef]
21. Jadhav, H.B.; Gogate, P.; Annapure, U. Process intensification of acidolysis reaction catalysed by enzymes for synthesis of designer lipids using sonication. *Chem. Eng. J.* **2022**, *428*, 131374. [CrossRef]
22. Mota, D.A.; Rajan, D.; Heinzl, G.C.; Osório, N.M.; Gominho, J.; Krause, L.C.; Soares, C.M.F.; Nampoothiri, K.M.; Sukumaran, R.K.; Ferreira-Dias, S. Production of low-calorie structured lipids from spent coffee grounds or olive pomace crude oils catalyzed by immobilized lipase in magnetic nanoparticles. *Bioresour. Technol.* **2020**, *307*, 123223. [CrossRef]
23. Nunes, P.A.; Pires-Cabral, P.; Ferreira-Dias, S. Production of olive oil enriched with medium chain fatty acids catalysed by commercial immobilised lipases. *Food Chem.* **2011**, *127*, 993–998. [CrossRef]
24. Nunes, P.A.; Pires-Cabral, P.; Guillén, M.; Valero, F.; Ferreira-Dias, S. Optimized production of MLM triacylglycerols catalyzed by immobilized heterologous Rhizopus oryzae lipase. *JAOCS J. Am. Oil Chem. Soc.* **2012**, *89*, 1287–1295. [CrossRef]
25. Sharma, R.; Chisti, Y.; Banerjee, U.C. Production, purification, characterization, and applications of lipases. *Biotechnol. Adv.* **2001**, *19*, 627–662. [CrossRef]
26. Simões, T.; Valero, F.; Tecelão, C.; Ferreira-Dias, S. Production of human milk fat substitutes catalyzed by a heterologous Rhizopus oryzae lipase and commercial lipases. *JAOCS J. Am. Oil Chem. Soc.* **2014**, *91*, 411–419. [CrossRef]
27. Tong, X.; Busk, P.K.; Lange, L. Characterization of a new sn-1,3-regioselective triacylglycerol lipase from Malbranchea cinnamomea. *Biotechnol. Appl. Biochem.* **2016**, *63*, 471–478. [CrossRef]
28. Domínguez de María, P.; Sinisterra, J.V.; Tsai, S.-W.; Alcántara, A.R. Carica papaya lipase (CPL): An emerging and versatile biocatalyst. *Biotechnol. Adv.* **2006**, *24*, 493–499. [CrossRef] [PubMed]
29. Guillaume, D.; Charrouf, Z. Argan oil and other argan products: Use in dermocosmetology. *Eur. J. Lipid Sci. Technol.* **2011**, *113*, 403–408. [CrossRef]
30. Haimeur, A.; Meskini, N.; Mimouni, V.; Ulmann, L.; Messaouri, H.; Pineau-Vincent, F.; Abouakil, N.; Tremblin, G. A comparative study on the effect of argan oil versus fish oil on risk factors for cardio-vascular disease in high-fat-fed rats. *Nutrition* **2019**, *57*, 32–39. [CrossRef]
31. Pagliuca, G.; Bozzi, C.; Gallo, F.R.; Multari, G.; Palazzino, G.; Porrà, R.; Panusa, A. Triacylglycerol "hand-shape profile" of Argan oil. Rapid and simple UHPLC-PDA-ESI-TOF/MS and HPTLC methods to detect counterfeit Argan oil and Argan-oil-based products. *J. Pharm. Biomed. Anal.* **2018**, *150*, 121–131. [CrossRef]
32. Charrouf, Z.; Guillaume, D. The argan oil project: Going from utopia to reality in 20 years. *OCL* **2018**, *25*, D209. [CrossRef]
33. Belcadi-Haloui, B.; Zekhnini, Z.; Hatimi, A. Comparative study of Argan and other edible oils stability under accelerated oxidation. *Indian J. Sci. Technol.* **2018**, *11*, 1–7. [CrossRef]

34. Charrouf, Z.; Guillaume, D. Argan oil: Occurrence, composition and impact on human health. *Eur. J. Lipid Sci. Technol.* **2008**, *110*, 632–636. [CrossRef]
35. Lundstedt, T.; Seifert, E.; Abramo, L.; Thelin, B.; Nyström, Å.; Pettersen, J.; Bergman, R. Experimental design and optimization. *Chemom. Intell. Lab. Syst.* **1998**, *42*, 3–40. [CrossRef]
36. Tecelão, C.; Silva, J.; Dubreucq, E.; Ribeiro, M.H.; Ferreira-Dias, S. Production of human milk fat substitutes enriched in omega-3 polyunsaturated fatty acids using immobilized commercial lipases and Candida parapsilosis lipase/acyltransferase. *J. Mol. Catal. B Enzym.* **2010**, *65*, 122–127. [CrossRef]
37. Duan, Z.Q.; Du, W.; Liu, D.H. The solvent influence on the positional selectivity of Novozym 435 during 1,3-diolein synthesis by esterication. *Bioresour. Technol.* **2010**, *101*, 2568–2571. [CrossRef]
38. Vázquez, L.; González, N.; Reglero, G.; Torres, C. Solvent-free lipase-catalyzed synthesis of diacylgycerols as low-calorie food ingredients. *Front. Bioeng. Biotechnol.* **2016**, *4*, 6. [CrossRef]
39. Cabrera, Z.; Fernandez-Lorente, G.; Fernandez-Lafuente, R.; Palomo, J.M.; Guisan, J.M. Novozym 435 displays very different selectivity compared to lipase from Candida antarctica B adsorbed on other hydrophobic supports. *J. Mol. Catal. B Enzym.* **2009**, *57*, 171–176. [CrossRef]
40. Turan, S.; Karabulut, I.; Vural, H. Effects of reaction parameters on the incorporation of caprylic acid into soybean oil for production of structured lipids. *J. Food Lipids* **2006**, *13*, 306–317. [CrossRef]
41. Kim, I.H.; Kim, H.; Lee, K.T.; Chung, S.H.; Ko, S.N. Lipase-catalyzed acidolysis of perilla oil with caprylic acid to produce structured lipids. *JAOCS J. Am. Oil Chem. Soc.* **2002**, *79*, 363–367. [CrossRef]
42. Sousa, V.; Campos, V.; Nunes, P.; Pires-Cabral, P. Incorporation of capric acid in pumpkin seed oil by *sn*-1,3 regioselective lipase-catalyzed acidolysis. *OCL* **2018**, *25*, A302. [CrossRef]
43. Abed, S.M.; Zou, X.; Ali, A.H.; Jin, Q.; Wang, X. Synthesis of 1,3-dioleoyl-2-arachidonoylglycerol-rich structured lipids by lipase-catalyzed acidolysis of microbial oil from Mortierella alpina. *Bioresour. Technol.* **2017**, *243*, 448–456. [CrossRef]
44. Stergiou, P.Y.; Foukis, A.; Filippou, M.; Koukouritaki, M.; Parapouli, M.; Theodorou, L.G.; Hatziloukas, E.; Afendra, A.; Pandey, A.; Papamichael, E.M. Advances in lipase-catalyzed esterification reactions. *Biotechnol. Adv.* **2013**, *31*, 1846–1859. [CrossRef]
45. Haaland, P.D. *Experimental Design in Biotechnology*; CRC Press: New York, NY, USA, 1989; ISBN 9781003065968.
46. Tecelão, C.; Rivera, I.; Sandoval, G.; Ferreira-Dias, S. Carica papaya latex: A low-cost biocatalyst for human milk fat substitutes production. *Eur. J. Lipid Sci. Technol.* **2012**, *114*, 266–276. [CrossRef]
47. Wang, S.; Meng, X.; Zhou, H.; Liu, Y.; Secundo, F.; Liu, Y. Enzyme stability and activity in non-aqueous reaction systems: A mini review. *Catalysts* **2016**, *6*, 32. [CrossRef]
48. Rodrigues, J.; Canet, A.; Rivera, I.; Osório, N.M.; Sandoval, G.; Valero, F.; Ferreira-Dias, S. Biodiesel production from crude Jatropha oil catalyzed by non-commercial immobilized heterologous Rhizopus oryzae and Carica papaya lipases. *Bioresour. Technol.* **2016**, *213*, 88–95. [CrossRef]
49. De Martini Soares, F.A.S.; Osório, N.M.; da Silva, R.C.; Gioielli, L.A.; Ferreira-Dias, S. Batch and continuous lipase-catalyzed interesterification of blends containing olive oil for trans-free margarines. *Eur. J. Lipid Sci. Technol.* **2013**, *115*, 413–428. [CrossRef]
50. Rodrigues, R.C.; Virgen-Ortíz, J.J.; dos Santos, J.C.S.; Berenguer-Murcia, Á.; Alcantara, A.R.; Barbosa, O.; Ortiz, C.; Fernandez-Lafuente, R. Immobilization of lipases on hydrophobic supports: Immobilization mechanism, advantages, problems, and solutions. *Biotechnol. Adv.* **2019**, *37*, 746–770. [CrossRef] [PubMed]
51. Xu, X. Production of specific-structured triacylglycerols by lipase-catalyzed reactions: A review. *Eur. J. Lipid Sci. Technol.* **2000**, *102*, 287–303. [CrossRef]

life

Article

Co-Processed Olive Oils with *Thymus mastichina* L.—New Product Optimization

Fátima Peres [1,2], Marta Roldão [2], Miguel Mourato [2], Luisa L. Martins [2] and Suzana Ferreira-Dias [2,*]

1 Instituto Politécnico de Castelo Branco, Escola Superior Agrária, 6000-909 Castelo Branco, Portugal; fperes@ipcb.pt
2 LEAF, Linking Landscape, Environment, Agriculture and Food, Instituto Superior de Agronomia, Universidade de Lisboa, 1349-017 Lisbon, Portugal; marta_roldao97@hotmail.com (M.R.); mmourato@isa.ulisboa.pt (M.M.); luisalouro@isa.ulisboa.pt (L.L.M.)
* Correspondence: suzanafdias@mail.telepac.pt

Abstract: Olive co-processing consists of the addition of ingredients either in the mill or in the malaxator. This technique allows selecting the type of olives, the ingredients with the greatest flavoring and bioactive potential, and the technological extraction conditions. A new product—a gourmet flavored oil—was developed by co-processing olives with *Thymus mastichina* L. The trials were performed using overripe fruits with low aroma potential (cv. 'Galega Vulgar'; ripening index 6.4). Experimental conditions were dictated by a central composite rotatable design (CCRD) as a function of thyme (0.4–4.6%, *w/w*) and water (8.3–19.7%, *w/w*) contents used in malaxation. A flavored oil was also obtained by adding 2.5% thyme during milling, followed by 14% water addition in the malaxator (central point conditions of CCRD). The chemical characterization of the raw materials, as well as the analysis of the flavored and unflavored oils, were performed (chemical quality criteria, sensory analysis, major fatty acid composition, and phenolic compounds). Considering chemical quality criteria, the flavored oils have the characteristics of "Virgin Olive Oil" (VOO), but they cannot have this classification due to legislation issues. Flavored oils obtained under optimized co-processing conditions (thyme concentrations > 3.5–4.0% and water contents varying from 14 to 18%) presented higher phenolic contents and biologic value than the non-flavored VOO. In flavored oils, thyme flavor was detected with high intensity, while the defect of "wet wood", perceived in VOO, was not detected. The flavored oil, obtained by *T. mastichina* addition in the mill, showed higher oxidative stability (19.03 h) than the VOO and the co-processed oil with thyme addition in the malaxator (14.07 h), even after six-month storage in the dark (16.6 vs. 10.3 h).

Keywords: co-extraction; flavored oil; response surface methodology; phenols; thyme

Citation: Peres, F.; Roldão, M.; Mourato, M.; Martins, L.L.; Ferreira-Dias, S. Co-Processed Olive Oils with *Thymus mastichina* L.—New Product Optimization. *Life* **2021**, *11*, 1048. https://doi.org/10.3390/life11101048

Academic Editor: Othmane Merah

Received: 29 August 2021
Accepted: 2 October 2021
Published: 6 October 2021

1. Introduction

Virgin olive oil is the oil extracted from the fruits of the olive tree (*Olea europaea* L.) using exclusively mechanical extraction techniques under conditions that will not affect the original composition of the oil. In the last decades, many research studies have shown that virgin olive oil has bioactive properties with impacts on health, particularly in preventing cardiovascular diseases, cancer, diabetes, and neurodegenerative diseases [1–4]. These benefits are due not only to its fatty acid profile, especially to the high content in oleic acid, but also to the presence of several bioactive compounds [1,5–7]. This evidence led the European Union, in 2006, to approve nutritional and health claims for virgin olive oil (VOO), to be included on the label. They concern monounsaturated fatty acids, oleic acid, and unsaturated acids, vitamin E, and more recently, in 2012, polyphenols in the olive oil [8,9].

Most of the Portuguese olive orchards in organic production are predominantly rainfed and very much based on autochthonous varieties, which are characterized by unique sensory properties. Combining these aspects with the consumers' trend towards products offering health benefits, the great challenge to promote and protect this type of olive orchard is related to the production of high-quality oils (extra virgin olive oil—EVOO) and novel added-value products. Labeling these products with nutrition and health claims is very important in terms of consumers' acceptance and choice. In recent years, innovation in olive oils has included flavoring with different ingredients, with the aim of improving their sensory and nutritional properties, as well as shelf life. Thus, this type of product is highly appreciated, especially among consumers outside the Mediterranean countries, reaching values above EUR 50 per liter. The added compounds may have several health benefits due to the presence of natural bioactive substances with antioxidant and/or antimicrobial properties and may contribute to increasing olive oil's resistance to oxidation [10]. Vegetables, aromatic herbs, fruits, nuts, essential oils, and spices are the most common ingredients used, added either as an infusion, ethanolic extracts of essential oils, or by co-processing [10–16].

Co-processing, also known as co-extraction, is an alternative method to infusion techniques, or to the addition of ethanolic extracts, to obtain flavored or enriched oils. It consists of the addition of ingredients, for example, fruits or aromatic plants, during milling or in the malaxation step of the olive oil extraction process. This technique allows for selecting both the type of olives (e.g., cultivar and ripening stage), the ingredient(s) with the greatest flavoring and/or bioactive potential, as well as the extraction conditions. Moreover, this method does not need the filtration step, conversely to when infusion is performed. For olive oils from organic farming, the production of co-processed oils can be an opportunity for differentiation, creating products mainly for the non-traditional consumer market. Although studies of co-processing are scarce, some show that using co-extraction with the addition of lemon, bergamot, rosemary, thyme, basil, and oregano causes positive sensory notes [10,13,17–19]. In addition, the use of citrus fruits or their peels in co-extraction increases the antioxidant activity of olive oil and its nutritional value [18,20,21]. The use of spices in co-malaxation increases the antioxidant activity of olive oils compared to the infusion technique [13]. Tomato by-products are also used in co-milling to enrich olive oils with lycopene [22].

Thymus mastichina L. (T. *mastichina*) is an Iberian endemic thyme used in the food industry as a condiment and herbal infusion and as a source of essential oil for the cosmetics industry [23,24]. It is also known as mastic thyme, Spanish marjoram, or white thyme. Traditionally, it was used for treating digestive, respiratory, and rheumatic disorders [25]. The major compounds in the essential oil of *T. mastichina* are 1,8-cineol (58.8–64.1%) and α-terpineol (5.6%) [26]. The main phenolic compounds identified in *T. mastichina* extracts are rosmarinic acid, methoxysalicylic acid, apigenin, kaempferol, luteolin, chlorogenic acid, cafeic acid, and derivatives of luteolin and apigenin [24,27]. Thus, the selection of this species of thyme for co-extraction trials is based on the use of an endogenous plant as well as on its potential, not only in terms of sensory attributes but also for improving the biological value and oxidative stability of enriched flavored oils. Therefore, the aim of this study is to develop a new value-added product, a gourmet flavored oil, based on co-processing overripe healthy 'Galega Vulgar' olives, with low intensity of fruity aroma and low amounts of bioactive phenolic compounds, and *Thymus mastichina* L. from organic farming. The effect of the addition of dried thyme, either in olive milling or in malaxation, is investigated. It is expected to obtain a gourmet oil with thyme flavor and, if possible, improved bioactivity and shelf-life. According to our knowledge, this is the first study on co-extraction of olives with this species of thyme and process optimization using response surface methodology.

2. Materials and Methods

2.1. Biological Material

Portuguese olive fruits of 'Galega Vulgar' cultivar used in the present study were produced in a rain-fed olive grove situated in the Beira Baixa region (39°50′ N, 7°42′ W), Portugal. 'Galega Vulgar' fruits were picked in January 2021, with a ripening index (RI) of 6.4 and an average weight of 2.5 ± 0.1 g per fruit with a very low water content (42.31% ± 0.21%), were used for the co-processing experiments with thyme addition in the malaxation operation. 'Galega Vulgar'olives with a RI of 6.2 were used for the co-processing trials with the addition of thyme in the mill. Dried *Thymus mastichina* L. was purchased from Ervas de Zoé, Ladoeiro, Portugal, and was produced according to organic farming (OF) guidelines.

2.2. Milling of Thymus Mastichina L. and Particle Size Classification

The dried plants of *T. mastichina* were submitted to milling (hammer mill PX-MFC90D from Kinematic, Switzerland, exit grid of 2 mm opening). Sieve analysis was performed to classify the particles according to their size, using five sieves of Tyler equivalent series (10, 28, 35, 60, and 140 mesh, equivalent to opening sieves of 1.68, 1, 0.42, 0.25, 0.106 mm, respectively)

2.3. Co-Processing

Flavored oils were obtained in a laboratory oil mill (Abencor analyzer; MC2 Ingenieria y Sistemas S.L., Seville, Spain), comprising a hammer mill, a malaxation unit, and a cylindrical bowl centrifuge. The olives (c.a. 10 kg) were crushed at 3000 rpm, using a 5 mm grid in the mill.

Co-processing experiments with thyme addition in the malaxator were performed following a central composite rotatable design (CCRD) as a function of the contents of thyme and water [28,29]. In each trial, 0.5 kg of olives was used. In this design, the five levels tested for thyme and water concentration were between 0.4 and 4.6% (w/w) and between 8.3 and 19.7% (w/w), respectively (Table 1). Water and thyme were added at the beginning of the malaxation step, carried out at 28–30 °C for 30 min. Paste centrifugation was performed at 3500 rpm for 1 min. After centrifugation, the water traces in the oil were removed with anhydrous sodium sulfate, which was removed by filtration through a cellulose filter (Whatman 41) [30]. After, the oils were collected in amber flasks and stored at 4 °C until analysis.

The combined effects of the concentrations of *T. mastichina* (*Tm*) and of water (*W*) on the oil extraction yield, total phenols, chlorophyll pigments, major fatty acid composition, as well as on chemical quality criteria parameters (acidity, peroxide value, and UV absorbances) of the extracted oils, were investigated by response surface methodology (RSM). RSM allows finding the optimal conditions with a smaller number of experiments than the conventional approach (one variable at a time, OVAT), with the same precision as OVAT and with the advantage of giving information about possible interactions between the variables, which is not possible following the OVAT approach [28,29].

Another set of experiments was carried out with a different batch of olives (RI = 6.2) with the addition of thyme in the mill, at the concentration of the central point of the CCRD (2.5%, w/w). Water was added to the malaxator at a concentration of 14% (conditions of the central point). The extraction yield was calculated as previously described [31]. Chemical and sensory analyses were performed within one week after extraction.

Table 1. CCRD followed in the experiments for co-processing as a function of the amounts of thyme and water added in malaxation. Experiment N° 14 (control) corresponds to the virgin olive oil obtained under the same extraction conditions as flavored oils.

Experimental Points	Experiment Number	[Thyme] Coded Value	[Water] Coded Value	[Thyme] (%) Decoded Values	[Water] (%) Decoded Values
Factorial points	1	−1	−1	1	10
	2	−1	1	1	18
	3	1	−1	4	10
	4	1	1	4	18
Star points	5	$-\sqrt{2}$	0	0.4	14
	6	$\sqrt{2}$	0	4.6	14
	7	0	$-\sqrt{2}$	2.5	8.3
	8	0	$\sqrt{2}$	2.5	19.7
Central points	9	0	0	2.5	14
	10	0	0	2.5	14
	11	0	0	2.5	14
	12	0	0	2.5	14
	13	0	0	2.5	14
Control	14	–	–	-	14

2.4. Shelf-Life Studies

Flavored and unflavoured oil were stored in amber glass bottles at 22–23 °C, in the dark, for 6-month shelf-life studies. These samples were analyzed as described in paragraph 2.6.

2.5. Proximate Analysis of Olives and Thyme

Samples (Galega olives after the milling process and thyme plants after milling (2.2)) were subjected to drying at 105 ± 2 °C (Selecta Drying Oven, JP Selecta, Barcelona, Spain) until constant weight, in order to evaluate their moisture content. Fat content was determined by extraction from dried samples in a Soxtec System HT2 Extraction unit (Tecator AB, Hoganas, Sweden), using petroleum ether as an organic solvent. For ash content assay, samples with an initial weight of around 0.5 g were placed in a muffle furnace at 550 °C (Heraeus Instruments, Hanau, Germany). After 24 h, the final weight of the samples was obtained, and the ash content was determined. For protein content assay, nitrogen content (N) was determined in dried samples according to the Kjeldahl method using 0.5 g of each sample (Velp Scientifica UDK 139, Usmate, Italy). The crude protein content was obtained using the conversion factor of 6.25. Mineral element content was determined as follows: approximately 0.4 g of each dried sample were ground in a mortar and weighed in a Teflon tube to which 3 mL of concentrated nitric acid (68%) and 10 mL of concentrated hydrochloric acid (37%) were added. The tubes were then placed in a digestion plate (DigiPrep MS, SCP Science, Quebec, QC, Canada) with a heating cycle to 95 °C that lasted 1 h and remained at 95 °C for 1 h more. After cooling at 20 °C, the samples were filtered into a 25 mL volumetric flask and the volume filled with distilled water. The quantification of the elements (Cu, Zn, Fe, Mn, Na, K, Ca, Mg, P, and S) was done with ICP-OES (Inductively coupled plasma optical emission spectroscopy, Thermo iCAP 7200, Thermo Fisher Scientific, Waltham, MA, USA). Appropriate standards were prepared from a stock solution (100 mg/L) containing the analyzed elements (SCP Science, PlasmaQUAL S22, Baie-D'Urfe, QC, Canada). Results were expressed as mg/kg dry matter (DW).

2.6. Chemical and Sensory Characterisation of Flavored Oils

Acidity (% free fatty acids, %FFA, expressed in oleic acid), peroxide value (PV), UV absorbances related with the formation of conjugated hydroperoxides (K_{232}) and secondary oxidation products (K_{270}), and the major fatty acids (C16:0, C18:0, C18:1, and C18:2) of oils were evaluated by NIR spectroscopy (MPA, Bruker Optics, Ettlingen, Germany). The calibration model B-Olive-Oil (Bruker Optics, Ettlingen, Germany) was used. Spectral information was obtained from olive oil samples, previously prepared obtained at 50 °C (IN600-A, Bruker, Ettlingen, Germany), in absorbance mode and at a wavelength of 1200 to 4000 cm^{-1}, with 8 cm^{-1} resolution and 32 scans.

Total phenols were extracted by liquid-liquid microextraction and evaluated by VIS spectroscopy (JASCO 7800, Jasco Inc., Tokyo, Japan) according to Pizarro et al. [32], as previously described [33]. Results were expressed as milligram of gallic acid equivalent per kilogram of oil (mg GAE/kg). Chlorophyll pigments were assayed in accordance with the IUPAC method proposed by Pokorný et al. [34] using a single beam spectrophotometer (Biochrom Libra S21, Biochrom Ltd., Cambridge, UK) to evaluate the absorbances of oils at 630, 670, and 710 nm against air. The results were expressed as mg pheophytin a/kg oil. All analyses were performed in triplicate.

Samples of flavored oils were also sensory evaluated by a trained panel [35]. A Quantitative Descriptive Analysis (QDA) was applied by using a profile sheet where an unstructured 10 cm length scale was used to mark the intensity of the descriptors [36]. This profile sheet mainly considered the positive attributes (e.g., orthonasal, retronasal, and gustative analysis) of the oils. If defects were found, it was necessary to identify them and quantify their intensity. Oxidative stability (OS) was measured using a Metrohm Rancimat model 670 (Metrohm, Herisau, Switzerland) (temperature of 120 °C; airflow of 20 L h^{-1}).

The profile of phenolic compounds was evaluated by high performance liquid chromatography (HPLC) according to the International Olive Council method with some modifications [37]. The phenolic compounds were recovered from the olive oil by liquid-liquid extraction using the procedure proposed by Pirisi et al. [38]. An Agilent 1100 HPLC system (Agilent, Santa Clara, CA, USA), consisting of a degasser, a quaternary pump, a column oven, an autosampler, and a UV detector, was used. The stationary phase was a Purospher C18 analytical column (150 mm × 3.9 mm × 4 µm). The mobile phase consisted of solutions of (A) 0.2% H_3PO_4 (v/v), (B) methanol, and (C) acetonitrile at a constant flow rate of 1 mL min^{-1}. The gradient program used was the one indicated by the IOC document [37]. The identification of phenolic compounds was carried out using standards for gallic acid, hydroxytyrosol, tyrosol, vanillic acid, caffeic acid, vanillin, *p*-coumaric acid, *o*-coumaric acid, cinnamic acid, luteolin, apigenin, rosmarinic acid, kaempferol, and pinoresinol. Syringic acid was used as an internal standard. Standards of hydroxytyrosol, tyrosol, vanillic acid, vanillin, caffeic acid, ferulic acid, *o*-coumaric, *p*-coumaric, apigenin, rosmarinic acid were purchased from Sigma-Aldrich (St. Louis, MO, USA), oleuropein and luteolin from Extrasynthese (Genay, France), and pinoresinol from TCI Europe (Zwijndrecht, Belgium).

2.7. Statistical Analysis

The obtained results of CCRD, as well as ANOVA (post hoc Tukey test was used; $p \leq 0.05$), were analyzed using the software Statistica, version 7, from Statsoft, Tulsa, OK, USA. The linear effects, as well as the quadratic effects of each factor (variable) tested (water and thyme concentrations) and of their linear interactions, on each response (extraction yield, total phenols, and chlorophyll pigments contents) were calculated. The significance of each effect was evaluated by analysis of variance. A response surface, described by a first or a second-order polynomial equation, was fitted to each set of experimental results. The first and second-order coefficients of these equations were generated by regression analysis. The goodness of fit of the polynomial models was evaluated by the coefficient of determination (R^2) and adjusted R^2 [28,29].

3. Results and Discussion

3.1. Proximate Analysis of Olives and Thyme

The results for moisture, fat, ash, protein, and mineral elements of olives and *T. mastichina* are presented in Table 2. The low moisture content of the olives highlights the need to perform the optimization of water addition in the malaxation trials, in combination with thyme addition. Galega olives and thyme fat contents (DW) are in accordance to Peres et al. [6] and to Barros et al. [39], respectively. In turn, ash and protein contents of thyme are higher than the values reported by Barros et al. [39]. For mineral composition, both materials have high contents of K (16.04 and 13.82 g/kg, in olives and thyme, respectively) and Ca (3.29 and 11.11 g/kg in olives and thyme, respectively). However, no references on the mineral content in *T. mastichin* a were found. The results obtained in our study are very different from those observed by Kassegn and Mekelle [40] for lemon thyme, where the microelement contents were 0.734, 1.630, 16.41, 0.106 mg kg^{-1} for P, Cu, Fe, and Mn, respectively. However, Kuçukbay and Kuyumcu [41] found values ranging from 6.5–14.90 mg/kg for Cu, 8.470–18.187 g/kg for K, and 8.383–25.570 g/kg for Ca, in other thyme species from Turkey, which are similar to the values obtained in the present study.

Table 2. Proximate analysis (DW) of olives and *Thymus mastichina* L.

Parameter	Unit	Olives	*T. mastichina*
Moisture	(%)	42.31± 0.21	6.30 ± 0.04
Fat	(%)	39.62± 0.82	3.08 ± 0.01
Ash	(%)	3.67 ± 0.59	5.35 ± 0.06
Protein	(%)	5.00 ± 0.62	7.79 ± 0.15
Cu	(mg/kg)	9.2 ± 1.3	10.1 ± 1.1
Zn		19.8 ± 3.0	83.2 ± 18.5
Fe		91.0 ± 6.4	218.5 ± 4.5
Mn		23.8 ± 1.3	292.7 ± 44.5
Na		154.8 ± 7.6	205.5 ± 4.8
K		16,040 ± 199	13,822 ± 1195
Ca		3288 ± 101	11,118 ± 464
Mg		472 ± 7	2178 ± 214
P		1199 ± 17	1958 ± 260
S		853 ± 33	1854 ± 107

3.2. Particle-Size Analysis of Milled Thyme

After milling, thyme particles, with dimensions smaller than 2 mm, were separated by fractions using a set of 5 sieves. Figure 1 shows that c.a. 70% of ground material is formed by particles with dimensions between 1 and 1.68 mm, corresponding to the fraction 10/18 mesh. Only 0.2% of the particles have dimensions higher than 1.68 and smaller than 2 mm, and 26.6% have dimensions smaller than 1 mm. Therefore, the milled thyme used in co-processing presented very homogeneous particles concerning their size. These particles are big enough to facilitate their removal from the oil together with the olive pomace by centrifugation in olive oil extraction plants, avoiding emulsion formation and oil loss.

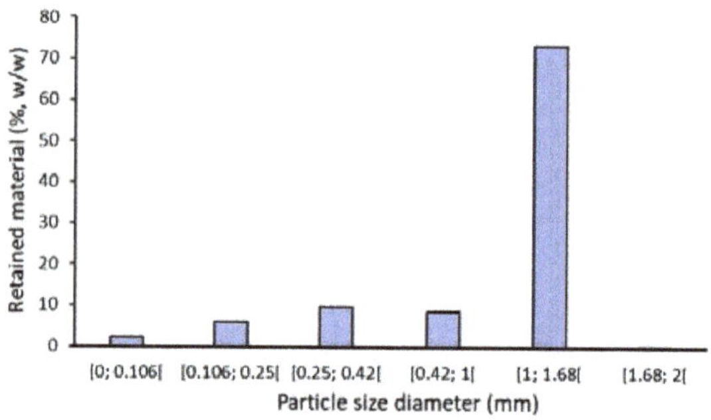

Figure 1. Histogram of different granulometric fractions of milled thyme.

3.3. Optimization of Co-Processing Conditions

In the 'Galega' cultivar, the oil yield obtained without thyme addition was 10.6%. In CCRD experiments, extraction yields in the presence of thyme varied from 5.1 to 10.7%. Thus, except for experiment 2 (1% thyme and 18% water), where 10.7% oil yield was obtained, the extraction yields obtained in the presence of thyme were lower than in the blank trial.

The statistical analysis of the oil yields showed that the addition of *T. mastichina* had a significant negative linear effect ($p = 0.014$) on oil extraction. It means that an increase in thyme concentration in the malaxation step will promote a decrease in extraction yield. This decrease may be ascribed to oil adsorption to the lignocellulosic material of thyme during malaxation operation, hindering mechanical oil extraction. A similar situation was referred by other authors, which was explained by an absorption/adsorption process [42,43]. For water addition, a positive linear effect was found ($p = 0.05$), meaning that an increase in water concentration will improve the extraction yield. A decrease in the viscosity of olive paste due to the presence of water may help oil extraction since diffusion is promoted. No significant quadratic effect of water and of the interaction effect of thyme and water addition was observed on oil yield. The negative quadratic effect of thyme ($p > 0.05$) was important enough to be retained in the response surface model. Its removal causes a great lack of fit of the model. In experimental design analysis, it is better to retain a "nonsignificant" effect ($p > 0.05$) in the model than to remove an important one [28]. Figure 2 shows the response surface fitted to oil extraction yield, Y, as a function of thyme (Tm; %, w/w) and water (W; %, w/w) concentrations. This is a convex surface described by the following second-order polynomial Equation (1):

$$Y = 6.93 + 0.335Tm - 0.239Tm^2 + 0.221W \quad (1)$$

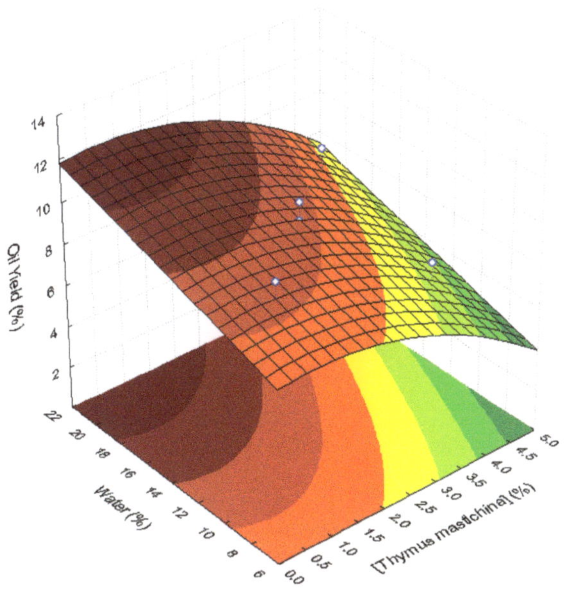

Figure 2. Response surface describing oil yield (%) as a function of water (%, w/w) and thyme concentration (%, w/w), added to 'Galega Vulgar' fruits during malaxation.

This model presents a good fit to the experimental points since it has a determination coefficient, R^2, of 0.69 and an adjusted determination coefficient (R^2_{Adj}) of 0.59. Therefore, 69% of the experimental results are explained by this model.

Higher oil yields were observed for higher water contents and lower thyme concentrations (Figure 2).

3.4. Characterization of Flavored Oils

The results of chemical quality criteria (acidity, PV, and UV absorbances) and major fatty acid composition (C16:0; C18:1; C18:2, and C18:3) of (i) the flavored oils obtained from CCDR trials, (ii) of one flavored oil obtained by co-processing with thyme addition during milling, and (iii) of the virgin olive oil obtained under the same extraction conditions without thyme addition, are presented in Table 3. The variation observed in chemical quality criteria of flavored oils, acidity (0.17−0.21%), PV (4.2−4.9 meq O_2 kg^{-1}), K_{232} (1.54−1.63), and K_{270} (0.11−0.15), is not significant. This shows that the co-processing with *T. mastichina* did not affect the quality of the flavored oils obtained. Moreover, a similar behavior was observed concerning the major fatty acid contents: palmitic acid varied from 11.54 to 12.22%; oleic acid ranged from 77.00 to 77.42%; linoleic acid varied from 4.56 to 5.06% and linolenic acid from 0.5 to 0.8%. In addition, flavored oils obtained by co-processing, either with thyme addition in the hammer mill or in the malaxator, have similar quality values and fatty acid composition. Other authors have stated that when the initial indices are relatively high, the obtained flavored oils may present quality parameters values above the legal limit for EVOO [17,44]. This was not observed in our study.

Table 3. Results of acidity (% oleic acid), peroxide value (meq O_2 kg^{-1}), UV absorbances (K_{232}, K_{270}), palmitic acid (%) (C16:0), oleic acid (%) (C18:1), linoleic acid (%) (C18:2), and linolenic acid (%) (C18:3) (conditions of each experiment are shown in Table 1; experiment 14 corresponds to the co-processed flavored oil obtained by adding thyme in the hammer mill; experiment 15 is the VOO extracted from the same fruits without thyme addition, i.e., the control).

Experiment	Acidity	PV	K_{232}	K_{270}	C16:0	C18:1	C18:2	C18:3
1	0.17	4.5	1.54	0.13	11.59	77.03	5.06	0.48
2	0.18	4.3	1.57	0.11	11.54	76.99	5.00	0.58
3	0.19	4.5	1.60	0.13	12.04	77.31	4.73	0.76
4	0.21	4.3	1.59	0.12	12.06	77.21	4.66	0.75
5	0.18	4.2	1.58	0.11	11.59	77.09	4.86	0.59
6	0.18	4.7	1.61	0.15	12.22	77.42	4.56	0.82
7	0.21	4.7	1.62	0.14	11.85	77.07	4.85	0.67
8	0.17	4.3	1.60	0.12	11.97	77.10	4.63	0.67
9	0.18	4.5	1.61	0.14	11.74	77.28	4.57	0.70
10	0.17	4.9	1.61	0.14	11.87	77.16	4.73	0.72
11	0.19	4.2	1.61	0.14	11.93	77.04	4.73	0.70
12	0.18	4.7	1.63	0.14	11.97	77.08	4.77	0.71
13	0.18	4.6	1.61	0.14	11.88	77.14	4.70	0.73
14 (hammer mill)	0.11	4.5	1.72	0.18	11.93	77.02	4.78	0.81
15 (VOO-control)	0.20	4.4	1.59	0.12	11.47	76.94	4.86	0.62

The amounts of total phenolic compounds (*TPH*) in flavored oils varied from 60.7 to 141.6 mg GAE/kg oil, while the VOO obtained without co-processing with thyme had a *TPH* of 71.4 ± 6.8 mg GAE/kg oil. In fact, not all the flavored oils obtained by co-processing with *T. mastichina* added in the malaxator presented higher amounts of phenolic compounds than the original VOO. The low content of *TPH* in VOO is explained by the high ripening stage of the olives used, together with a relatively low *TPH* of Galega VOO, when compared to VOO from other cultivars at the same RI [6,45]. The flavored oil obtained under the conditions of the central point (Table 1: 2.5% thyme and 14% water) showed a *TPH* of 115.4 ± 4.9 mg GAE/kg oil, corresponding to a *TPH* increase of 61.6% when compared with the non-flavored VOO. The co-processed flavored oil obtained by the addition of thyme in the mill, under the conditions of the central point, presented a *TPH* of 189.66 ± 6.8 mg GAE/kg oil, which represents an increase of 75% with respect to the original VOO. Thus, phenolic compounds extraction seems to be more efficient by co-processing with *T. mastichina* added in the milling than in the malaxation operation. These results show the importance of performing optimization trials before the addition of the flavoring agent.

Data analysis of CCRD showed that the content of phenolic compounds in flavored oils linearly increased with thyme concentration (p = 0.004) and with water added in the malaxator (p = 0.06). A significant negative quadratic effect of water concentration (p = 0.045), indicating a convex quadratic response as a function of this effect, was also found. No significant effects of thyme at quadratic level or of the interaction thyme x water were found. Therefore, a convex response surface, described by the following second-order polynomial, can be fitted to *TPH* (mg GAE/kg) as a function thyme (*Tm*; %, *w*/*w*) and water (*W*; %, *w*/*w*) concentrations (Figure 3a), Equation (2):

$$TPH = -98.19 + 12.91Tm + 22.99W - 0.732W^2 \quad (2)$$

The good fit of the model is demonstrated by high values of both R^2 (0.78) and R^2_{Adj} (0.71).

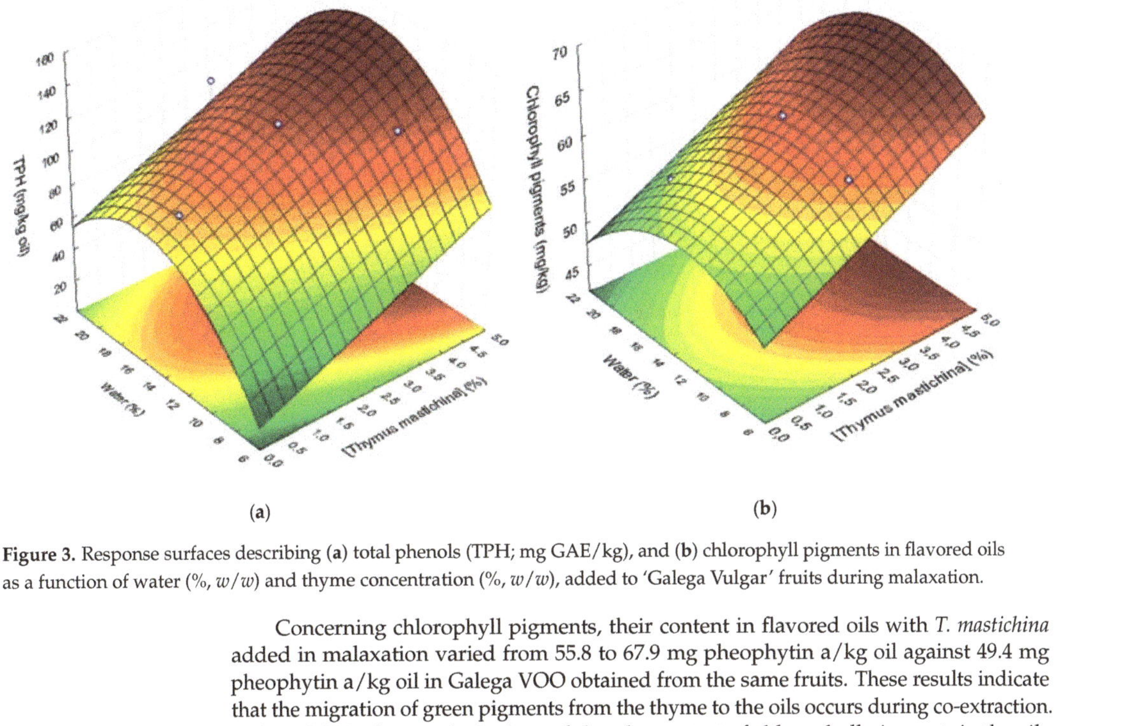

(**a**) (**b**)

Figure 3. Response surfaces describing (**a**) total phenols (TPH; mg GAE/kg), and (**b**) chlorophyll pigments in flavored oils as a function of water (%, *w*/*w*) and thyme concentration (%, *w*/*w*), added to 'Galega Vulgar' fruits during malaxation.

Concerning chlorophyll pigments, their content in flavored oils with *T. mastichina* added in malaxation varied from 55.8 to 67.9 mg pheophytin a/kg oil against 49.4 mg pheophytin a/kg oil in Galega VOO obtained from the same fruits. These results indicate that the migration of green pigments from the thyme to the oils occurs during co-extraction. In fact, CCRD data analysis showed that the content of chlorophyll pigments in the oils significantly depends on the thyme amount ($p = 10^{-6}$) added during malaxation, increasing with it. The water added in the malaxator showed to have a negative linear effect (p = 0.06) and a negative quadratic effect (p = 0.008) on green pigment extraction to the oils. Thus, chlorophyll pigments in flavored oils can be described by a convex surface as a function of thyme and water concentrations in pastes during malaxation (Figure 3b). This response surface is given by the following second-order polynomial equation where the amount of chlorophyll pigments (*CP*; mg/kg oil) is a function of thyme content (*Tm*; %, *w*/*w*) and water (*W*; %, *w*/*w*) concentrations, Equation (3):

$$CP = 39.65 + 2.70Tm + 2.39W - 0.092W^2 \quad (3)$$

This model has a very good fit to the experimental results (R^2 = 0.96; R^2_{Adj} = 0.94).

Figure 3 shows that both *TPH* and chlorophyll pigments are described by similar shape convex response surfaces. The highest *TPH* values are obtained for co-processing with thyme concentrations higher than c.a. 3.5−4.0% and water contents in the range c.a. 14−18%. The highest pigment contents are also obtained with thyme concentrations higher than c.a. 3.5% and water content of the pastes between 10 and 18%. Conversely, the highest extraction yields are observed for thyme contents lower than 2.5% and when high water contents are used (>18%) (Figure 2).

According to several authors, the addition of thyme (T. *vulgaris*) in olive oil increases the *TPH* [46,47], especially when the co-extraction technique is used [19]. The presence of high amounts of water in olive pastes during malaxation has been related to the loss of hydrophilic phenols in the water phase [1,48]. Therefore, in our study, the production of high-quality flavored oils enriched with bioactive phenolic compounds extracted from *T. mastichina* was obtained under conditions that will conduct to lower oil extraction yields. The use of RSM showed to be a useful tool to find the best operation conditions for flavored oil production by co-extraction. The evaluation of the profiles of oil yield, *TPH*, and chlorophylls in oils obtained under different co-extraction conditions (water and thyme concentrations) was only possible via the visual observation of each response surface fitted to each data set. The optimal co-extraction conditions were chosen from the information shown in these response surfaces.

3.5. Shelf-Life Studies: Quality, Phenol Composition, Sensory Analysis, and Oxidative Stability

The VOO and flavored oil samples obtained by co-processing with *T. mastichina*, added either in the malaxation or in milling operations, under the conditions of the central point of the CCRD (Table 1), were submitted to shelf-life studies. After 6 months of storage in the dark at 22−23 °C, VOO and flavored oils were assayed for their chemical parameters and sensory properties. No significant differences were observed among VOO and flavored oils in terms of quality parameters and fatty acid composition (Table 4). Moreover, for each quality parameter, no significant differences were observed between the initial values and those obtained after 6 months of storage, except for PV, which showed around a 66% increase in stored oils. This indicates that, after 6 months of storage in the dark, oil oxidation was at the initial induction stage of hydroperoxide formation. Along this storage period, *TPH* and chlorophyll pigments decreased 8.6 and 7.6%, respectively.

Table 4. Results of acidity (% oleic acid), peroxide value (meq O_2kg^{-1}), UV absorbances (K_{232}, K_{270}), palmitic acid (%) (C16:0), oleic acid (%) (C18:1) and linoleic acid (%) (C18:2) of the co-processed flavored oils obtained by adding thyme in the malaxator (experiments 9–13. corresponding to the central point of Table 1) or in the hammer mill (experiment 14); and of the virgin olive oil extracted from the same fruits without thyme addition (experiment 15), after 6 months storage at 22−23 °C in the dark.

Experiment	Acidity	PV	K_{232}	K_{270}	C16:0	C18:1	C18:2	C18:3	TPH
9	0.21	7.39	1.74	0.18	11.90	2.60	4.78	0.61	120.08
10	0.20	7.07	1.74	0.16	11.84	2.58	4.92	0.58	108.19
11	0.23	6.89	1.76	0.16	11.88	2.60	4.94	0.60	97.33
12	0.22	7.65	1.78	0.17	11.95	2.63	4.96	0.63	107.13
13	0.22	7.58	1.77	0.16	11.83	2.57	4.97	0.9	99.28
14 (hammer mill)	0.25	6.7	1.78	0.19	11.83	2.62	4.62	0.70	158.60
15 (VOO-control)	0.25	7.5	1.78	0.15	11.59	2.45	5.14	0.50	69.42

In order to identify the phenols that migrated from the T. *mastichina* to the oils, increasing their bioactivity, an HPLC profile at 280 nm (all phenol compounds identified by standards), 320 nm (*p*-coumaric acid), and 360 nm (flavonoids) was performed. The chromatographic phenolic profiles at 280 nm of thyme-flavored oils, compared to unflavored VOO, are shown in Figure 4.

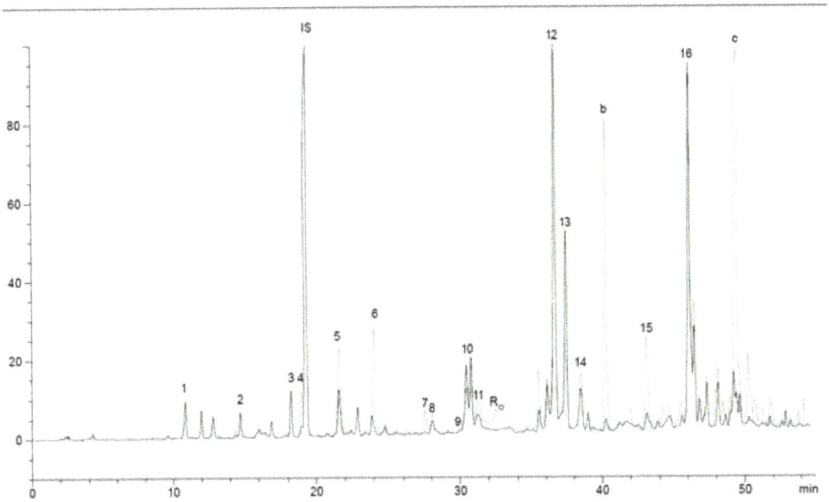

Figure 4. HPLC-RP-VWD 280 nm phenolic profile of the unflavored VOO (continuous line) and flavored oil (dotted line). (1—hydroxytyrosol; 2—tyrosol; 3—vanilic acid; IS—Internal standard (syringic acid) 4—caffeic acid; 5—vanillin; 6—*p*-coumaric acid; 7—unknown; 8—luteolin-7-glucosid; 9—*o*-coumaric acid; 10—3,4-DHPEA-EDA (literature); 11—oleuropein; Ro- Rosmarinic acid; 12—pinoresinol; 13—cinnamic acid 14—luteolin; 15—apigenin; 16—*p*-HPEA-EA (literature); b and c—unknown).

In our study, as a result of the co-extraction with *T. mastichina*, none of the VOO phenolic compounds disappeared in flavored oils. An increase in some phenolic compounds, already present in VOO, as well as the presence of new compounds, were observed. In fact, flavored oils showed an increase in vanillin (peak 5), *p*-coumaric acid (peak 6), luteolin (peak 14), apigenin (peak 15), an unidentified peak at a retention time (RT) at 40 min (peak b: non-flavonoid, not detected at 360 nm), and an unidentified peak at RT 50 min (peak c) already present in VOO.

As reported by other studies, the main phenolic compounds identified in *T. mastichina* extracts are rosmarinic acid, methoxysalicylic acid, apigenin, kaempferol, luteolin, chlorogenic acid, cafeic acid, and derivatives of luteolin and apigenin [24,27]. The HPLC profiles of standards show that rosmarinic acid has an RT of 31.67 min and higher absorbance at 320 nm. In flavored samples of our study, rosmarinic acid showed not to be the main phenolic compound transferred from the thyme to the oil (Figure 4). Chlorogenic acid (RT = 16.97 min), caffeic acid (RT = 18.8 min), and kaempferol (RT = 43.13 min), referred to in the literature as present in *Thymus mastichina*, were not transferred to the oil.

Peaks 10 and 16 are referred in the literature as 3,4-DHPEA-EDA (oleacein or dialdehydic form of elenolic acid linked to hydroxytyrosol) and *p*-HPEA-EA (ligstroside aglycone, aldehyde, and hydroxylic form), respectively [37]. Comparing the profile of unflavored Galega VOO obtained in our study with those from the literature [6], higher amounts of 3,4-DHPEA-EDA and *p*-HPEA-EA were observed in VOO from fruits with low ripening indices, which are characterized by high phenol content and flavor intensity.

The sensory analysis was also performed for the original VOO and flavored oils obtained under the conditions of the central point of the CCRD, either by the addition of *T. mastichina* in the mill or in the malaxator. These oils were sensory evaluated immediately after extraction and after 6 months of storage (Figure 5). After extraction, Galega VOO exhibited a very low (<2) intensity of ripe fruity (orthonasal and retronasal evaluation), bitterness, and pungency. Moreover, defects ascribed to frostbitten olives (wet wood) were detected by assessors with an intensity lower than 3.5. Therefore, despite chemical quality criteria corresponding to the extra virgin olive oil category (Table 4), due to the identification of this defect, the oil can no longer be classified as EVOO but falls in the category of "virgin olive oil". The presence of this sensory defect has already been reported in VOO from fruits after weather-related hazards like frost [49,50]. After 6 months of storage in the dark at 22–23 °C, the defect was perceived with similar intensity; a decrease in the intensity of ripe fruity (orthonasal and retronasal) was registered in VOO, while bitter and pungent attributes were no longer detected.

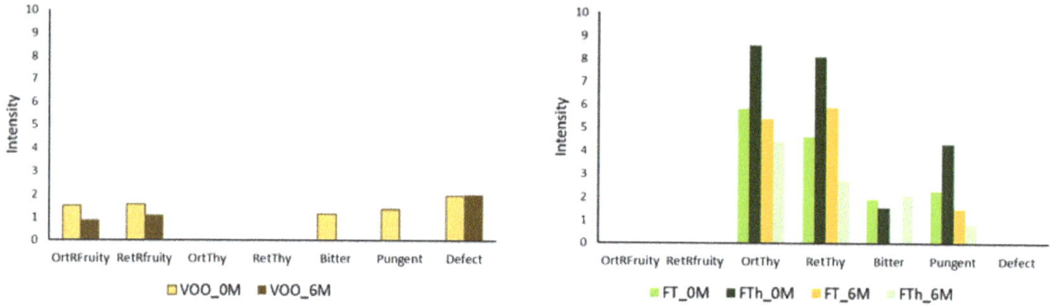

(a) (b)

Figure 5. Sensory evaluation of (**a**) the VOO without storage (VOO_0M) and after 6 months storage at 22–23 °C in the dark (VOO_6 M), and of (**b**) thyme flavored oils in the malaxation step (FT_0M) and in the hammer mill (FTh_0M) without storage and after 6 months storage at 22–23 °C in the dark (FT_6 M and FTh_6M, respectively). OrtRFruity—orthonasal ripe olive fruity; RetRfruity—retronasal ripe fruity; OrtThy—orthonasal thyme; RetThy—retronasal thyme.

In the flavored oils with *Thymus mastichina*, the defect of "wet wood" perceived in VOO was not detected. A thyme flavor (orthonasal and rethonasal), with very high intensity (≥8) for the hammer mill co-processed oils, and of intensity 5–6, in malaxator co-processed flavored oils, was registered. Moreover, the olive fruity (orthonasal and rethonasal) notes, characteristic of EVOO, were almost absent in thyme-flavored oils. These results indicate that strong changes in the volatile fraction occurred due to the co-processing of olives with thyme, as previously reported by others [51].

After 6 months of storage, no defects were detected in flavored oils. However, a decrease in thyme flavor was observed in both flavored oils, being more pronounced in the oil flavored by the addition of thyme in milling. The bitter taste had similar intensity in both fresh flavored oils. After 6 months of storage, the bitter taste was absent in the flavored oil with thyme addition during malaxation, while its initial intensity was maintained in the flavored oil with thyme addition in milling. Concerning the initial intensity of pungency, it was higher in the flavored oil obtained by thyme addition in milling, but, at the end of the storage experiment, the intensity of this attribute was similar in both flavored oils.

In conclusion, the flavored oil obtained by co-processing with *T. mastichina* in the mill showed higher initial intensity scores for positive attributes but, after 6 months of storage, its thyme flavor was less intense than that of the oil obtained by co-processing with thyme addition in the malaxator.

In Figure 6, the results of oxidative stability (OS) of VOO and flavored oils by co-extraction with *T. mastichina* added either in the mill or in the malaxator are presented. Co-processing with the addition of thyme in a hammer mill (FTh) improves the OS of the flavored oils (14.07 h in VOO vs. 19.03 h in flavored oil), while no significant differences were found between VOO and thyme flavored oils in the malaxation step (FT) (14.30 h). After six months of storage, the oils showed a significant decrease in OS. No differences were observed between the VOO and the flavored oil with the addition of thyme in malaxation (c.a. 10.3 h), but the OS of the flavored oil obtained by co-extraction with *T. mastichina* added in the mill was higher, with a value of 16.06 h (corresponding to a 15.6% reduction after 6 months of storage).

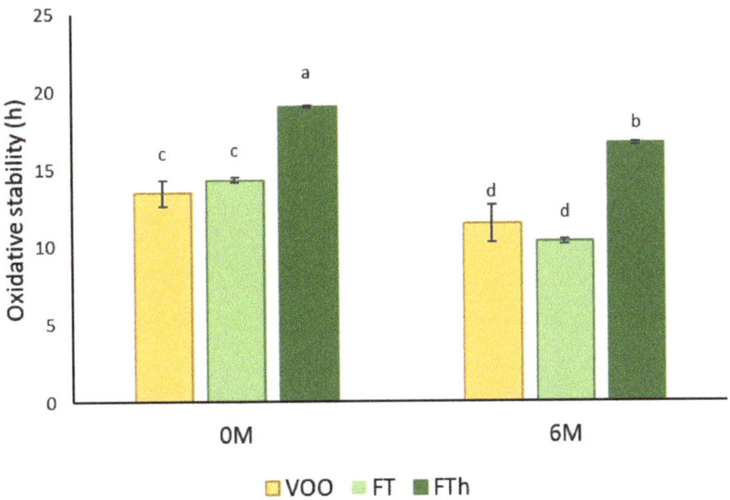

Figure 6. Oxidative stability (hour) of VOO and thyme flavored oils obtained by co-extraction in the malaxation (FT) or in the mill (FTh) at time 0 (0M) and after 6 months of storage (6M) at 22–23 °C in the dark (different letters of each bar (a, b, c, d) indicate significant differences in the oxidative stability of the oils at $p < 0.05$ according to Tukey test).

In Rubió et al. [52], the addition of *Thymus zygis* extracts significantly improved the oxidative stability of oils when compared to VOO, while Issaoui et al. [47] only detected a nonsignificant increase in OS of flavored oils. Conversely, Issaoui et al. [53] found that the oxidative stability of oils decreased with the addition of thyme, regardless of the concentration added (20 to 80 g/kg).

Depending on several factors, such as the type, the level, and the method of incorporation of the flavoring agent, the initial properties of the olive oil, as well as the storage conditions, results in increasing or decreasing trends in OS have been reported [10,16,54–56]. Our results confirm that direct malaxation of the olive paste with thyme or its addition in the hammer mill is a technique that is easy to carry out. Moreover, it does not include any additional unit operation to the conventional olive oil extraction process and is faster than infusion. In addition, both flavoring techniques are conventional green processes that do not require the use of any organic solvents [13].

4. Conclusions

Consumer's attitude towards olive oil is changing. Their search for different flavored oils that match several food pairings is an opportunity for the olive oil sector. Thus, the research on developing adequate technologies to produce these novel oils, presenting adequate sensory properties, improved biological value, and shelf-life stability, is of utmost importance for the olive oil industry. Although different techniques can be used, some of them consisting of the incorporation of the flavoring agent in the olive oil may have a deleterious effect on nutritional and shelf-life characteristics. This study showed the feasibility of sustainably producing new high-value products obtained with oils from autochthonous olive cultivars using natural resources. Olive oils obtained from overripe olives, presenting a slight aroma and low amounts of bioactive phenolic compounds, can be highly valorized by co-extraction with thyme. The optimization of co-extraction and storage studies is a good way to evaluate the main effects of the addition of thyme. The co-processing of ripe olives with *T. mastichina* added, either in the milling or in the malaxation operations, showed to be a feasible and easy technique to obtain flavored oils with intense thyme sensory notes and without sensory defects that were present in the VOO extracted from the same olives. Considering chemical quality criteria, these flavored oils have the characteristics of VOO, but they cannot have this classification due to legislation issues. Under optimized co-processing conditions, flavored oils presented higher phenolic contents, and higher biologic value, than the non-flavored VOO. The flavored oil obtained by co-processing with *T. mastichina* addition in the mill showed higher oxidative stability than the VOO and the co-processed oil with thyme addition in the malaxator, even after six-month storage in the dark.

Author Contributions: Conceptualization, F.P. and S.F.-D.; methodology, F.P., M.M., L.L.M., S.F.-D.; formal analysis, M.R., F.P., M.M.; investigation, M.R., F.P.; data curation, F.P., S.F.-D., M.R.; writing—original draft preparation, F.P., M.R., S.F.-D.; writing—review and editing, F.P., M.M., L.L.M., S.F.-D.; supervision, F.P., S.F.-D.; project administration, F.P. All authors have read and agreed to the published version of the manuscript.

Funding: This work was funded by the Fundação para a Ciência e a Tecnologia (FCT), Portugal, through the research unit LEAF—Linking Landscape, Environment, Agriculture and Food Research Centre (UIDB/04129/2020).

Institutional Review Board Statement: Not applicable.

Informed Consent Statement: Not applicable.

Data Availability Statement: Data are available upon request to the authors.

Acknowledgments: The authors thank Conceição Vitorino and Cecilia Gouveia, for the excellent technical assistance, Paula Gonçalves for the help in some of the experimental determinations, the Biotech Plant Lab of Beira Interior, Castelo Branco, Portugal, and the LEAF—Linking Landscape, Environment, Agriculture, and Food Research Centre (UIDP/04129/2020).

Conflicts of Interest: The authors declare no conflict of interest.

References

1. Servili, M.; Selvaggini, R.; Esposto, S.; Taticchia, A.; Montedoro, G.F.; Morozzib, G. Health and sensory properties of virgin olive oil hydrophilic phenols: Agronomic and technological aspects of production that affect their occurrence in the oil. *J. Chromatogr. A* **2004**, *1054*, 113–127. [CrossRef]
2. Cornwell, D.G.; Ma, J. Nutritional Benefit of Olive Oil: The Biological Effects of Hydroxytyrosol and Its Arylating Quinone Adducts. *J. Agric. Food Chem.* **2008**, *56*, 8774–8786. [CrossRef]
3. Cicerale, S.; Lucas, L.; Keast, R. Biological Activities of Phenolic Compounds Present in Virgin Olive Oil. *Int. J. Mol. Sci.* **2010**, *11*, 458–479. [CrossRef] [PubMed]
4. Psaltopoulou, T.; Kosti, R.I.; Haidopoulos, D.; Dimopoulos, M.; Panagiotakos, D.B. Olive oil intake is inversely related to cancer prevalence: A systematic review and a meta-analysis of 13,800 patients and 23,340 controls in 19 observational studies. *Lipids Health Dis.* **2011**, *10*, 127. [CrossRef]
5. Bendini, A.; Cerretani, L.; Carrasco-Pancorbo, A.; Caravaca, A.M.G.; Segura-Carretero, A.; Fernández-Gutiérrez, A.; Lercker, G. Phenolic Molecules in Virgin Olive Oils: A Survey of Their Sensory Properties, Health Effects, Antioxidant Activity and Analytical Methods. An Overview of the Last Decade. *Molecules* **2007**, *12*, 1679–1719. [CrossRef]
6. Peres, F.; Martins, L.L.; Mourato, M.; Vitorino, C.; Antunes, P.; Ferreira-Dias, S. Phenolic compounds of 'Galega Vulgar' and 'Cobrançosa' olive oils along early ripening stages. *Food Chem.* **2016**, *211*, 51–58. [CrossRef]
7. Wani, T.A.; Masoodi, F.A.; Gani, A.; Baba, W.N.; Rahmanian, N.; Akhter, R.; Wani, I.A.; Ahmad, M. Olive oil and its principal bioactive compound: Hydroxytyrosol—A review of the recent literature. *Trends Food Sci. Technol.* **2018**, *77*, 77–90. [CrossRef]
8. Regulation (EC) No 1924/2006 of the European Parliament and of the Council of 20 December 2006 on Nutrition and Health Claims Made on Foods. 2006. Available online: http://data.europa.eu/eli/reg/2006/1924/oj (accessed on 4 January 2021).
9. Commission Regulation (EU) No 432/2012 of 16 May 2012 Establishing a List of Permitted Health Claims Made on Foods, Other than Those Referring to the Reduction of Disease Risk and to Children's Development and Health. 2012. Available online: http://data.europa.eu/eli/reg/2012/432/oj (accessed on 4 January 2021).
10. Baiano, A.; Terracone, C.; Gambacorta, G.; Notte, E.L. Changes in Quality Indices, Phenolic Content and Antioxidant Activity of Flavored Olive Oils during Storage. *J. Am. Oil Chem. Soc.* **2009**, *86*, 1083. [CrossRef]
11. Mannina, L.; D'Imperio, M.; Gobbino, M.; D'Amico, I.; Casini, A.; Emanuele, M.C.; Sobolev, A.P. Nuclear magnetic resonance study of flavoured olive oils. *Flavour Fragr. J.* **2012**, *27*, 250–259. [CrossRef]
12. Caporaso, N.; Paduano, A.; Nicoletti, G.; Sacchi, R. Capsaicinoids, antioxidant activity, and volatile compounds in olive oil flavored with dried chili pepper (*Capsicum annuum*). *Eur. J. Lipid Sci. Technol.* **2013**, *115*, 1434–1442. [CrossRef]

13. Caponio, F.; Durante, V.; Varva, G.; Silletti, R.; Previtali, M.A.; Viggiani, I.; Squeo, G.; Summo, C.; Pasqualone, A.; Gomes, T.; et al. Effect of infusion of spices into the oil vs. combined malaxation of olive paste and spices on quality of naturally flavoured virgin olive oils. *Food Chem.* **2016**, *202*, 221–228. [CrossRef]
14. Reboredo-Rodríguez, P.; Figueiredo-González, M.; González-Barreiro, C.; Simal-Gándara, J.; Salvador, M.D.; Cancho-Grande, B.; Fregapane, G. State of the Art on Functional Virgin Olive Oils Enriched with Bioactive Compounds and Their Properties. *Int. J. Mol. Sci.* **2017**, *18*, 668. [CrossRef] [PubMed]
15. González-Gamallo, S.; Salvador, M.D.; Fregapane, G. Design and Characteristics of Novel Sensory and Nutritionally Oriented Olive, Seed, and Nut Virgin Oils' Blendings. *Eur. J. Lipid Sci. Technol.* **2021**, *123*, 2100008. [CrossRef]
16. Cherif, M.; Rodrigues, N.; Veloso, A.C.A.; Zaghdoudi, K.l.; Pereira, J.A.; Peres, A.M. Kinetic-thermodynamic study of the oxidative stability of Arbequina olive oils flavored with lemon verbena essential oil. *LWT* **2021**, *140*, 110711. [CrossRef]
17. Sacchi, R.; Della Medaglia, D.; Paduano, A.; Caporaso, N.; Genovese, A. Characterisation of lemon-flavoured olive oils. *LWT Food Sci. Technol.* **2017**, *79*, 326–332. [CrossRef]
18. Cerretani, L.; Bendini, A.; Poerio, A.; Toschi, T.G. Citric acid as co-adjuvant: Improvement of the antioxidant activity of edible olive oils. *Agro Food Ind. Hi-Tech* **2008**, *19*, 64–66.
19. Clodoveo, M.L.; Dipalmo, T.; Crupi, P.; Durante, V.; Pesce, V.; Maiellaro, I.; Lovece, A.; Mercurio, A.; Laghezza, A.; Corbo, F.; et al. Comparison Between Different Flavored Olive Oil Production Techniques: Healthy Value and Process Efficiency. *Plant Foods Hum. Nutr.* **2016**, *71*, 81–87. [CrossRef]
20. Ascrizzi, R.; Taglieri, I.; Sgherri, C.; Flamini, G.; Macaluso, M.; Sanmartin, C.; Venturi, F.; Quartacci, M.F.; Pistelli, L.; Zinnai, A. Nutraceutical Oils Produced by Olives and Citrus Peel of Tuscany Varieties as Sources of Functional Ingredients. *Molecules* **2019**, *24*, 65. [CrossRef]
21. Ergönül, P.G.; Sánchez, S. Evaluation of polycyclic aromatic hydrocarbons content in different types of olive and olive pomace oils produced in Turkey and Spain. *Eur. J. Lipid Sci. Technol.* **2013**, *115*, 1078–1084. [CrossRef]
22. Bendini, A.; Di Lecce, G.; Valli, E.; Barbieri, S.; Tesini, F.; Gallina Toschi, T. Olive oil enriched in lycopene from tomato by-product through a co-milling process. *Int. J. Food Sci. Nutr.* **2015**, *66*, 371–377. [CrossRef]
23. Méndez-Tovar, I.; Sponza, S.; Asensio-S-Manzanera, M.C.; Novak, J. Contribution of the main polyphenols of *Thymus mastichina subsp. mastichina* to its antioxidant properties. *Ind. Crop. Prod.* **2015**, *66*, 291–298. [CrossRef]
24. Taghouti, M.; Martins-Gomes, C.; Schäfer, J.; Santos, J.A.; Bunzel, M.; Nunes, F.M.; Silva, A.M. Chemical Characterization and Bioactivity of Extracts from *Thymus mastichina*: A Thymus with a Distinct Salvianolic Acid Composition. *Antioxidants* **2020**, *9*, 34. [CrossRef]
25. Pardo-de-Santayana, M.; Tardío, J.; Blanco, E.; Carvalho, A.M.; Lastra, J.J.; San Miguel, E.; Morales, R. Traditional knowledge of wild edible plants used in the northwest of the Iberian Peninsula (Spain and Portugal): A comparative study. *J. Ethnobiol. Ethnomedicine* **2007**, *3*, 27. [CrossRef] [PubMed]
26. Cunha, A.P.; Roque, O.R. *Especiarias e Plantas Condimentares. Origem, Composição e Utilizações*; Fundação Calouste Gulbenkian: Lisboa, Portugal, 2015.
27. Delgado, T.; Marinero, P.; Asensio, S.; Manzanera, M.C.; Asensio, C.; Herrero, B.; Pereira, J.A.; Ramalhosa, E. Antioxidant activity of twenty wild Spanish *Thymus mastichina* L. populations and its relation with their chemical composition. *LWT-Food Sci. Technol.* **2014**, *57*, 412–418. [CrossRef]
28. Haaland, P.D. *Experimental Design in Biotechnology*; Marcel Dekker Inc.: New York, NY, USA, 1989.
29. Montgomery, D.C. *Design and analysis of Experiments*, 3rd ed.; John Wiley & Sons: New York, NY, USA, 2017.
30. ISO 661. *Animal and Vegetable Fats and Oils—Preparation of Test Sample*; ISO: Geneve, Switzerland, 2003.
31. Peres, F.; Martins, L.L.; Ferreira-Dias, S. Laboratory-scale optimization of olive oil extraction: Simultaneous addition of enzymes and microtalc improves the yield. *Eur. J. Lipid Sci. Technol.* **2014**, *116*, 1054–1062. [CrossRef]
32. Pizarro, M.L.; Becerra, M.; Sayago, A.; Beltrán, M.; Beltrán, R. Comparison of Different Extraction Methods to Determine Phenolic Compounds in Virgin Olive Oil. *Food Anal. Methods* **2013**, *6*, 123–132. [CrossRef]
33. Peres, F.; Talhinhas, P.; Afonso, H.; Alegre, H.; Oliveira, H.; Ferreira-Dias, S. Olive Oils from Fruits Infected with Different Anthracnose Pathogens Show Sensory Defects Earlier Than Chemical Degradation. *Agronomy* **2021**, *11*, 1041. [CrossRef]
34. Pokorny, J.; Kalinová, L.; Dysseler, P. Determination of Chlorophyll pigments in Crude Vegetable Oils. *Pure Appl. Chem.* **1995**, *67*, 1781–1787.
35. Official Journal of the European Union. *Commission Delegated Regulation (EU) No 2019/1604 of 27 September 2019 amending Regulation (EEC) No 2568/91 on the Characteristics of Olive Oil and Olive-Residue Oil and on the Relevant Methods of Analysis*; Official Journal of the European Union: Brussels, Belgium, 2019; Volume L250, pp. 14–48. Available online: https://eur-lex.europa.eu/legal-content/EN/TXT/PDF/?uri=CELEX:32019R1604&from=EN (accessed on 4 January 2021).
36. Cerretani, L.; Salvador, M.D.; Bendini, A.; Fregapane, G. Relationship Between Sensory Evaluation Performed by Italian and Spanish Official Panels and Volatile and Phenolic Profiles of Virgin Olive Oils. *Chem. Percept.* **2008**, *1*, 258–267. [CrossRef]
37. IOC. Determination of Biophenols in Olive Oils by HPLC. COI/T20/DOC.29. 2009. Available online: https://www.oelea.de/downloads/COI-T20-DOC-29-2009-DETERMINATION-OF-BIOPHENOLS-IN-OLIVE-OILS-BY-HPLC.pdf (accessed on 4 January 2021).
38. Pirisi, F.M.; Cabras, P.; Cao, C.F.; Migliorini, M.; Muggelli, M. Phenolic Compounds in Virgin Olive Oil. 2. Reappraisal of the Extraction, HPLC Separation, and Quantification Procedures. *J. Agric. Food Chem.* **2000**, *48*, 1191–1196. [CrossRef] [PubMed]

39. Barros, L.; Carvalho, A.M.; Ferreira, I.C.F.R. From famine plants to tasty and fragrant spices: Three Lamiaceae of general dietary relevance in traditional cuisine of Trás-os-Montes (Portugal). *LWT-Food Sci. Technol.* **2011**, *44*, 543–548. [CrossRef]
40. Kassegn, H.H.; Mekelle, E.P. Inorganic and phytochemical content analysis of the wild abyssinian thyme spice. *Food Sci. Qua. Manage.* **2016**, 82–85.
41. Kuçukbay, F.; Kuyumcu, E. Determination of trace element contents of Thymus species from Turkey. *Turk. J. Chem.* **2010**, *34*, 911–920.
42. Squeo, G.; Difonzo, G.; Summo, C.; Crecchio, C.; Caponio, F. Study of the influence of technological coadjuvants on enzyme activities and phenolic and volatile compounds in virgin olive oil by a response surface methodology approach. *LWT* **2020**, *133*, 109887. [CrossRef]
43. Sadkaoui, A.; Jiménez, A.; Pacheco, R.; Beltrán, G. Micronized natural talc affects the proteins and pectic cell wall polysaccharides during "Hojiblanca" virgin olive oil extraction. *Eur. J. Lipid Sci. Technol.* **2017**, *119*, 1600039. [CrossRef]
44. Gambacorta, G.; Faccia, M.; Pati, S.; Lamacchia, C.; Baiano, A.; La Notte, E. Changes in the chemical and sensorial profile of extravirgin olive oils flavored with herbes and spices during storage. *J. Food Lipids* **2007**, *14*, 202–215. [CrossRef]
45. Rodrigues, N.; Casal, S.; Pinho, T.; Peres, A.M.; Bento, A.; Baptista, P.; Pereira, J.A. Ancient olive trees as a source of olive oils rich in phenolic compounds. *Food Chem.* **2018**. [CrossRef] [PubMed]
46. Ayadi, M.A.; Grati-Kamoun, N.; Attia, H. Physico-chemical change and heat stability of extra virgin olive oils flavoured by selected Tunisian aromatic plants. *Food Chem. Toxicol.* **2009**, *47*, 2613–2619. [CrossRef] [PubMed]
47. Issaoui, M.; Flamini, G.; Souid, S.; Bendini, A.; Barbieri, S.; Gharbi, I.; Toschi, T.G.; Cioni, P.L.; Hammami, M. How the Addition of Spices and Herbs to Virgin Olive Oil to Produce Flavored Oils Affects Consumer Acceptance. *Nat. Prod. Commun.* **2016**, *11*. [CrossRef]
48. Diamantakos, P.; Giannara, T.; Skarkou, M.; Melliou, E.; Magiatis, P. Influence of Harvest Time and Malaxation Conditions on the Concentration of Individual Phenols in Extra Virgin Olive Oil Related to Its Healthy Properties. *Molecules* **2020**, *25*, 2449. [CrossRef]
49. Guillaume, C.; Ravetti, L.; Gwyn, S. Characterisation of Phenolic Compounds in Oils Produced from Frosted Olives. *J. Am. Oil Chem. Soc.* **2010**, *87*, 247–254. [CrossRef]
50. García-Vico, L.; García-Rodríguez, R.; Sanz, C.; Pérez, A.G. Biochemical aspects of olive freezing-damage: Impact on the phenolic and volatile profiles of virgin olive oil. *LWT* **2017**, *86*, 240–246. [CrossRef]
51. Baiano, A.; Previtali, M.A.; Viggiani, I.; Varva, G.; Squeo, G.; Paradiso, V.M.; Summo, C.; Gomes, T.; Caponio, F. As oil blending affects physical, chemical, and sensory characteristics of flavoured olive oils. *Eur. Food Res. Technol.* **2016**, *242*, 1693–1708. [CrossRef]
52. Rubió, L.; Motilva, M.-J.; Macià, A.; Ramo, T.; Romero, M.-P. Development of a Phenol-Enriched Olive Oil with Both Its Own Phenolic Compounds and Complementary Phenols from Thyme. *J. Agric. Food Chem.* **2012**, *60*, 3105–3112. [CrossRef] [PubMed]
53. Issaoui, M.; Flamini, G.; Hajaij, M.E.; Cioni, P.L.; Hammami, M. Oxidative Evolution of Virgin and Flavored Olive Oils under Thermo-oxidation Processes. *J. Am. Oil Chem. Soc.* **2011**, *88*, 1339–1350. [CrossRef]
54. Bobiano, M.; Rodrigues, N.; Madureira, M.; Dias, L.G.; Veloso, A.C.A.; Pereira, J.A.; Peres, A.M. Unmasking Sensory Defects of Olive Oils Flavored with Basil and Oregano Using an Electronic Tongue-Chemometric Tool. *J. Am. Oil Chem. Soc.* **2019**, *96*, 751–760. [CrossRef]
55. Khemakhem, I.; Yaiche, C.; Ayadi, M.A.; Bouaziz, M. Impact of Aromatization by Citrus limetta and Citrus sinensis Peels on Olive Oil Quality, Chemical Composition and Heat Stability. *J. Am. Oil Chem. Soc.* **2015**, *92*, 701–708. [CrossRef]
56. Taoudiat, A.; Djenane, D.; Ferhat, Z.; Spigno, G. The effect of *Laurus nobilis* L. essential oil and different packaging systems on the photo-oxidative stability of Chemlal extra-virgin olive oil. *J. Food Sci. Technol.* **2018**, *55*, 4212–4222. [CrossRef]

life

Article

Improving the Biological Value of Olive and Soybean Oil Blends with Olive Leaf Extract Obtained by Ultrasound-Assisted Extraction towards the Preparation of a Sauce Product

Mohammad Amin Aliyari and Karamatollah Rezaei *

Department of Food Science, Engineering and Technology, Karaj Campus, College of Agriculture and Natural Resources, University of Tehran, Karaj 31587-77871, Iran; amin.aliyari@ut.ac.ir
* Correspondence: krezaee@ut.ac.ir

Abstract: French sauce from different blends of soybean and olive oils was prepared and the oxidative stability of the optimum sauce sample, enriched with various amounts of olive leaf polyphenolic extract (OLE) (obtained via ultrasound-assisted extraction), was investigated over 90 days of storage. The microbiological and sensory properties of the samples containing the optimum amounts of OLE, as a substitution for synthetic preservatives, were studied. According to the results, the addition of olive oil at higher levels (75% and 100%) could affect the physicochemical properties of the sauce as compared to the control sample. It was also found that the addition of olive oil (up to 50%) would not significantly impact the sauce properties. Regarding the OLE enrichment in the samples, it was found that high levels of OLE could improve the oxidative stability of the samples. It was also found that OLE could be used as a preservative instead of commercial ones. Overall, this study suggests the potential use of olive oil and olive leaf extract in the preparation of French sauce to boost its nutritional value and its stability.

Keywords: emulsion; oxidative stability; microbiological criteria; phenolic compounds; physicochemical characteristics

Citation: Aliyari, M.A.; Rezaei, K. Improving the Biological Value of Olive and Soybean Oil Blends with Olive Leaf Extract Obtained by Ultrasound-Assisted Extraction towards the Preparation of a Sauce Product. *Life* **2021**, *11*, 974. https://doi.org/10.3390/life11090974

Academic Editors: Suzana Ferreira-Dias, Xuebing Xu and Fátima Peres

Received: 24 August 2021
Accepted: 12 September 2021
Published: 15 September 2021

1. Introduction

Sauces and salad dressings are oil in water (O/W) emulsions that have drawn the attention of consumers for their daily diets. Sauces are used to improve the flavor, taste, and appearance of many food products and among them French sauce (FS) is one of the most well-known ones [1]. Generally, these types of products are called cold-sauces as any amount of heating treatment would result in an emulsion breakdown [2]. Increasing demand for health-promoting products has led salads to get special attention as healthy food products. This, in turn, has promoted the use of sauces as one of the main commodities in our daily diets. FS has been a matter of interest among scientists. For instance, Rasmussen et al. [3] showed the potential use of honey as a good substitute for EDTA and commercial sweeteners. In an another study, de Melo et al. [4] evaluated the stability and nutritional index of FS samples with the use of mannoprotein from spent brewer's yeast. A typical FS usually consists of egg yolk, vinegar, oil, and tomato paste. In addition, several ingredients, such as paprika, mustard, and garlic powder, are needed as the main participants for sauces' specific taste and flavor. Moreover, texturizing agents are normally used to form the consistency of the sauce [5]. In sauce formulations, the type and nature of the oil used are very important. In sauces and dressings, oil phase plays a vital role in their stability and general appearance. Additionally, the nutritional value of sauces can be considered as a direct function of their oil phase. Using oils with high amounts of saturated fatty acids can easily lead to several health problems such as obesity, cardiovascular disease, and cancer [6].

Olive oil is a source of beneficial components, such as polyphenolics, hydrocarbons, sterols, pigments, volatile compounds, and vitamins [7]. Olive oil is relatively resistant to oxidation because of its noticeable amount of monounsaturated fatty acids (MUFAs) as well as the presence of natural antioxidants such as α-tocopherol and phenolic compounds, such as tyrosol, hydroxytyrosol, and others [8]. Among the MUFAs, oleic acid contributes 58–83% of the fatty acids in olive oil, which has been proven to provide multiple beneficial impacts such as anti-inflammatory and anticancer effects [9]. Because of wide health benefits for olive oil, it can be used as a new oil phase in the preparation of O/W model emulsions such as FS and mayonnaise. In general, finding a new oil phase with functional and health properties which does not alter the physical and sensorial properties of the final product, has been a matter of interest among researchers for many years [10].

One of the most important factors in sauces and salad dressings is their stability. Because of their relatively high content of oil and perishable ingredients such as egg yolk, they are likely to spoil quickly [6]. Thus, in order to preserve and maintain them for a long time, the use of preservatives is usual in the sauce industry. The presence of commercial preservatives, such as sorbate and benzoate salts, is a major issue in these types of products; as they can cause several health problems such as allergic reactions or different cancers [11]. Using natural preservatives instead of synthetic ones is growing rapidly. For example, Rafiee et al. [12] used nanoliposomes which contained polyphenolic compounds of pistachio green hull as a natural preservative agent in a mayonnaise formulation. In an another study, Bruzewicz et al. [13] used baicalin (from baical skullcap root) to prepare mayonnaise samples to improve their shelf life.

Nowadays, food waste products are considered as excellent sources of bioactive compounds such as polyphenolics, protein residues, and fibers. Therefore, the possibility of using them in a wide variety of food formulations and also reducing the environmental burden is of high interest for both food technologists and environmental activists. Using food sector by-products provides major benefits to the larger world population by supplying additional food sources at lower costs [14–16]. Olive leaves, which are the by-products of olive processing industries, are considered to be a rich sources of polyphenolic compounds, which have been used for food, medicine, and cosmetic applications [17]. Oleuropein, which is classified as a phenolic bitter compound, is found abundantly in olive leaves and offers several health benefits, such as cardio-protection, neuro-protection, antiaging, anti-inflammatory, anticancer, antioxidant, and antimicrobial activities [18,19]. Tyrosol is another polyphenolic compound that is found in olive leaves at high quantities. Şahin and Bilgin [20] reported that the antioxidant activity of hydroxyl tyrosol is ten times as high as that of green tea polyphenolics. Because of the antimicrobial properties of olive leaves, they could be a promising natural ingredient to help enhance the shelf life of emulsion-based products. Additionally, due to the high polyphenolic content of olive leaves, they can also improve the overall nutritional value of products. Thus, it is crucial that the nutritional value of these products is assessed [21].

Physicochemical as well as oxidative and microbiological criteria are the most practical methods for analyzing emulsion-based products (mainly sauces) in terms of their nutritional value and stabilities. Considering the above points, the objectives of this study are to find the optimum amount of olive oil in combination with soybean oil (which is widely used in sauce industries) in the formulation of FS and to further apply olive leaf extract (OLE) obtained by ultrasound-assisted extraction (UAE) to improve the shelf life of FS. The possibility of using OLE instead of commercial synthetic preservatives (benzoate and sorbate salts) are also evaluated.

2. Materials and Methods

2.1. Materials

Egg yolk, sugar, salt, vinegar, citric acid, tomato paste, paprika, garlic, and mustard powders were purchased from a local supermarket in Tehran, Iran. Soybean oil, xanthan and guar gums were provided by Behrouz Nik Industrial Company (Karaj, Alborz

Province, Iran). Virgin olive oil without any additive was purchased from Massoud Oil Company (city of Roudbar, Guilan province of Iran). Olive leaves were harvested from olive trees in Karaj, Iran in mid-June from the Koroneiki cultivar. All chemicals were of analytical grade and obtained from Merck Chemical Company (Darmstadt, Germany). Commercial FS was purchased from a local supermarket in Tehran, Iran and was from Bijan Industrial Company.

2.2. Chemical Analysis of Soybean and Olive Oils

Basic parameters of olive and soybean oils (acid value (AV), peroxide value (PV), iodine value (IV), and saponification value (SV)) were measured according to AOCS official methods [22]. The total phenolic contents (TPC) of the oils were measured according to Sousa et al. [23]. The chlorophyll content (ChC) and carotenoid content (CaC)) were determined using the method reported by Issaoui et al. [24]. Table 1 shows the basic parameters of oils used in this study.

Table 1. Chemical analysis of soybean and olive oils (shown as SBO and OO, respectively) used in the experiments.

	AV	PV	SV	IV	TPC	ChC	CaC
SBO	0.1 ± 0.0	2.8 ± 0.6	195 ± 2	134 ± 1	4 ± 0.3	0.001 ± 0.000	0.025 ± 0.000
OO	1.5 ± 0.3	18.7 ± 0.2	191 ± 3	79 ± 2	178 ± 8	7.41 ± 0.00	2.16 ± 0.00

AV: acid value in mg NaOH/kg oil, PV: Peroxide value in meq O_2/kg oil, SV: Saponification value in mg KOH/g oil, IV: Iodine value in g I_2/100 g oil, TPC: Total phenolic content in mg GAE/kg oil, ChC: chlorophyll content in mg pheophytin, CaC: carotenoid content in mg lutein.

2.3. Preparation and Characterization of Olive Leaf Extract

To obtain olive leaf extract (OLE), an ultrasound-assisted extraction technique (UAE) was applied according to the procedure outlined by Difonzo et al. [17]. At first, 25.0 g of olive leaves were placed in an oven at 120 °C until the moisture content of leaves reached below 1%. Then, the leaves were powdered in a mixer (IKA-WERKE M20, IKA Company, Staufen, Germany). After that, distilled water was added to the powdered leaves in the ratio of 1:20 (*w*/*v*, respectively). Then, the leaves and water blend were placed in an ultrasonic bath at 40 kHz (Universal Ultrasonic Cleaner, DSA 100-SK2, Fuzhou, China) and exposed to ultrasound for 90 min at 35 ± 5 °C. Then, the mixture was centrifuged (Hettich-Universal 320, Hettich Company, Tuttlingen, Germany) (25 °C, 30 min, and 4000× *g*). At the end, the supernatant was carefully separated and lyophilized. The obtained and lyophilized OLE was then stored at −20 °C for the rest of experiments.

The method reported by Şahin and Şamlı [25] was used with minor modifications to measure the total phenolic content (TPC) of OLE. The absorbance of the samples was measured at 765 nm. The results were expressed as mg Gallic acid equivalents (GAE) per gram of the dried olive leaves (mg GAE/g dried olive leaf). Gallic acid at different concentrations (0.170–0.150 mg/mL) was used to plot the calibration curve. A regression coefficient of 0.9983 was obtained for the calibration curve. The TPC of the obtained extract was determined as 27 ± 0.6 (mg GAE/g dry leaf).

2.4. Preparation of French Sauce Samples

The following proportions of the compounds (*w*/*w*) were used for the preparation of FS according to the method of Mizani et al. [1] with some modifications: egg yolk (20%), vinegar (10%), mustard powder (2.0%), garlic powder (0.16%), paprika powder (0.4%), tomato paste (4.0%), citric acid (0.1%), sugar (6.0%), salt (2.0%), xanthan gum (0.64%), guar gum (0.26%), sodium benzoate (0.063%), potassium sorbate (0.012%), water (16.4%), oil (38%). Olive and soybean oils were used at five different ratios (0%, 25%, 50%, 75%, and 100%, *w*/*w*). The sample containing 100% soybean oil and 0% olive oil was used as the control sample. All ingredients were mixed at a controlled temperature of 25 °C. For preparation of the samples, first, egg yolk was stirred to produce a considerable amount of foam. Then, vinegar, mustard powder, and citric acid were added and completely dissolved. Then, tomato paste and other ingredients except for gums were inserted and completely mixed. After that, xanthan and guar gums were added to make a proper texture. Finally, oil and water were added while mixing (from slow to fast) the components to generate a suitable emulsion structure in the samples. Samples were then maintained at 25 °C before the experiments.

2.5. Emulsion Stability

Emulsion stabilities of FS samples were measured using the method reported by Mun et al. [26] with slight modifications. Briefly, Falcon tubes containing 15 g of sauce samples were placed in a water bath (80 °C, 30 min). Then, the samples were cooled down for ~10 min, followed by centrifugation at 3000× *g* for 10 min. After centrifugation, the oil layer was carefully removed. Finally, the samples were weighed without the oil layer. The emulsion stability of the samples was determined using Equation (1):

$$\text{Emulsion Stability (\%)} = \frac{F_1}{F_0} \times 100 \quad (1)$$

where F_0 and F_1 are the sample weights before and after centrifugation, respectively.

2.6. Color Measurements

Color parameters (a*, b*, L*) were recorded at 25 °C using a Chroma Meter (CR-400, Konica Minolta Sensing, Tokyo, Japan). The Chroma Meter was placed at a distance of about 10 cm from each sample and the image was taken. After that, the data related to each color parameter appeared on the screen. All parameters were measured in triplicate. Images from different areas of each sample were taken.

2.7. Texture Profile Analysis

Using a texture analyzer TA-XT2 (Stable Micro Systems, Godalming, UK), firmness and consistency levels of the sauce samples were measured following the procedure described by Patil and Benjakul [27] at 25 °C. Briefly, samples were transferred to the containers up to the 125-mL level and a disc with a 35-mm diameter was used for compression. The speed was fixed at 1 mm/sec, until reaching a depth of 40 mm. To measure firmness and consistency, the force-time curve was used, in which the maximum point of the force curve and the area of the curve were used as the firmness and consistency values for the samples successively.

2.8. Rheological Properties

A rheometer (Physica MCR 301, Anton Paar, Graz, Austria) was used to determine the rheological properties of FS samples with parallel geometry (diameter: 25 mm, gap: 1 mm) using the probe PP 40/S (Anton Paar, Graz, Austria). First, the strain sweep from 0.01% to 100% at a constant frequency of 1.0 Hz was accomplished to determine the linear viscoelastic range. Then, the frequency sweep test in the 0.1–100 Hz frequency range was obtained under a constant strain of 0.5% and at 25 °C.

2.9. pH Determination

A 5% (w/v) sample solution was prepared and the pH values of the samples were measured using a pH meter (Crison-GLP 22, Crison Instruments, Alella, Spain) at 25 °C. All tests were carried out in triplicate.

2.10. Characterization of Enriched French Sauce Containing 50:50 (v/v) Olive and Soybean Oils and Different Amounts of Olive Leaf Extract

For further analysis, sauce samples were made using 50% olive oil and 50% soybean oil in the formulations (according to the obtained results, this blend of oils was selected as the optimum sample, so it was used for the rest of the experiments). To observe the effects of pure OLE on the oxidative parameters, commercial preservatives were not used in this stage. OLE was added to sauce samples at different levels (500, 1000, 1500, and 2000 mg/kg). A sample without the addition of OLE was used as the control. Sauce oxidative stability tests were then carried out every 15 days for a 3-month storage at 25 °C.

Oxidative Stability Evaluation

The AV, PV and *p*-anisidine values (PAV) were measured according to the official recommended methods of AOCS [22].

The total oxidation value (Totox) of each sample was determined using Equation (2):

$$\text{Totox value} = (2 \times \text{PV}) + \text{PAV} \quad (2)$$

All the oxidative stability tests were carried out under the daylight conditions.

2.11. Use of OLE in the Sauce Samples

In the last part of this study, the optimum amount of OLE was selected and the microbiological activities of a sauce sample with the optimum amount of OLE (FS-OLE) and without the addition of commercial preservatives, and a sample with commercial preservatives and without OLE, were measured. Sensory evaluations were also carried out on those samples as well as with a commercial product.

2.11.1. Microbiological Control

Total plate counts for mesophilic aerobic bacteria, yeast and molds, and the presence of *E. coli* and lactic acid bacteria of two FS samples with the substitution of commercial preservatives/OLE, were determined in seven stages (at the beginning and every 15 days up to three months of storage). For each analysis, first, 10 g of FS sample was weighed in a test tube and homogenized with peptone water (90 mL). Then, suitable dilutions of each suspension (1.0 mL) were poured and spread on the total-count plates and selective agar plates (Merck, Darmstadt, Germany). The total plate counts of aerobic mesophilic bacteria were carried out by means of a standard plate count agar and incubated at 30 °C for 72 h. Yeast and molds were measured via YGC (yeast extract glucose chloramphenicol) agars, which were incubated at 25 °C for 6 days. In order to determine the presence of *E. coli*, 1 mL of each sample's suspension was added to 9 mL of lauryl sulfate broth (LSB) and incubated at 37 °C for 48 h. Acid-tolerant microorganisms were identified using orange serum agar (OSA) at 45 °C, in which OSA was added to 1 mL of sample and after mixing it was left to solidify at room temperature, after which, it was incubated at 30 °C for 5 days.

2.11.2. Sensory Analysis

Sensory analysis of three samples (with and without OLE and a commercial product) were performed according to the method explained by Liu, Xu, and Guo [28]. Briefly, a 9- point hedonic scale was used (1 = the poorest and 9 = the best). The characteristics of samples (appearance, odor, texture, color, taste, and overall acceptance) were evaluated by 25 panelists (21–30 years of age). Prior to the sensory evaluation, all panelists were familiar with FS as part of their own experiences. Bread and water were provided between testing the samples to reset the remaining taste from the previous samples.

2.12. Statistical Analysis

Statistical analyses among different experiments were carried out using the SPSS program for Windows (version 25.0, SPSS Inc., Chicago, IL, USA). The evaluation of the differences among the treatments were performed applying a one-way analysis of variance (ANOVA) with Duncan's test at $p < 0.05$.

3. Results and Discussion

3.1. Emulsion Stability

Emulsion stability is an important factor in emulsion-based systems. This factor is related to some phenomena, such as coalescence, flocculation, and Ostwald ripening [26]. In sauces and salad dressings, proteins can make a layer around oil droplets. The generated layer by proteins can play a protective role against the accumulation of oil droplets in one area [29]. The results of the emulsion stability of samples are shown in Table 2. According to the results, all samples showed a very high emulsion stability (99.4–99.9%) at day 0 of production. As sauces and salad dressings contain high amounts of oil in their formulations, particles cannot move easily through their matrix, thus the stability of these products are usually high [30]. The stability levels of the samples were slightly decreased as the ratio of olive oil increased. The sample containing 100% olive oil (100OO) showed the lowest stability (99.4%). However, the control sample (100SO) showed the highest stability level (99.9%). From a statistical point of view, addition of olive oil up to 50% did not cause a meaningful difference from the control sample both on day 0 and on day 90 of the storage time. Olive oil generally contains less PUFAs than soybean oil. The partial reduction in the stability of the emulsion with an increase in olive oil is due to the oil's natural fatty acid profile. Based on Magnusson et al. [31], a positive correlation was obtained between the amount of unsaturated fatty acids and the emulsion stability. There was a decreasing trend in the emulsion stabilities of the samples during the storage time, but the trend was much more obvious in the samples containing 75% and 100% olive oil (75OO and 100OO, respectively). This decreasing trend could be due to the sensitivity of saturated fatty acids to shear stress, so their crystalline structure can be damaged more easily [32].

Table 2. Emulsion Stability (%) of Sauce samples prepared at different ratios of olive and soybean oils (100SO: samples with 100% soybean oil, 25OO: samples with 75% soybean oil and 25% olive oil, 50OO: samples with 50% soybean oil and 50% olive oil, 75OO: samples with 25% soybean oil and 75% olive oil, and 100OO: samples with 100% olive oil).

Storage Period (Day)	Sample				
	100SO	25OO	50OO	75OO	100OO
0	99.9 ± 0.1 ^{Aa}	99.9 ± 0.1 ^{Aa}	99.9 ± 0.1 ^{Aa}	99.6 ± 0.1 ^{Ab}	99.4 ± 0.0 ^{Ac}
15	99.9 ± 0.0 ^{Aa}	99.7 ± 0.2 ^{ABa}	99.8 ± 0.1 ^{Aa}	99.1 ± 0.1 ^{Ab}	98.9 ± 0.1 ^{Ab}
30	99.8 ± 0.0 ^{Aa}	99.5 ± 0.1 ^{Bab}	99.5 ± 0.1 ^{ABab}	98.1 ± 0.8 ^{Bb}	98.8 ± 0.1 ^{Ac}
45	99.7 ± 0.1 ^{ABa}	99.5 ± 0.0 ^{Ba}	99.2 ± 0.2 ^{Ba}	96.9 ± 0.1 ^{Cb}	97.81 ± 0.7 ^{Bc}
60	99.6 ± 0.2 ^{Ba}	98.8 ± 0.0 ^{Cb}	98.4 ± 0.1 ^{Cb}	96.7 ± 0.7 ^{Cc}	95.8 ± 0.1 ^{Cd}
75	98.4 ± 0.1 ^{Ca}	98.5 ± 0.1 ^{CDb}	98.1 ± 0.1 ^{Cc}	96.7 ± 0.1 ^{Cd}	95.5 ± 0.5 ^{Ce}
90	98.4 ± 0.3 ^{Ca}	98.3 ± 0.3 ^{Da}	98.1 ± 0.8 ^{Ca}	96.6 ± 0.7 ^{Cb}	95.4 ± 0.5 ^{Cc}

The results are given as means ± SD, $n = 3$. a, b, c, d, e: In each row, values with the same letter (lower case) are not significantly different ($p > 0.05$). A, B, C, D: In each column, values with the same letter (upper case) are not significantly different ($p > 0.05$).

3.2. pH Analysis

pH is one of the most important parameters in the evaluation of sauces and salad dressings. A high pH value (usually more than 4.2) usually causes serious damage such as growing pathogenic microorganisms (e.g., *Staphylococcus aureus*). Constant pH monitoring is one of the most critical factors in preparing new formulations of emulsion-based products [33]. The results of pH tests are shown in Table 3. Regarding the obtained results, the initial pH values (day 0) of the samples did not show significant differences. However, they decreased during the storage period of 90 days. Sample 100SO showed the lowest pH value (3.36), while the sample 75OO showed the highest pH of 3.97. Similar to the findings of the current study, Worrasinchai et al. [33] reported a declining trend of pH over 64 days of storage in mayonnaise samples. The reduction in pH during the storage time could be related to the breakdown of ester groups and generation of acidic groups [34]. Moreover, the growth and activity of some microorganisms, mainly lactic acid bacteria, can cause a drop in the pH [33].

Table 3. pH values of French sauce samples prepared at different ratios of olive and soybean oils (100SO: samples with 100% soybean oil, 25OO: samples with 75% soybean oil and 25% olive oil, 50OO: samples with 50% soybean oil and 50% olive oil, 75OO: samples with 25% soybean oil and 75% olive oil, and 100OO: samples with 100% olive oil).

Storage Period (Day)	Sample				
	100SO	25OO	50OO	75OO	100OO
0	4.10 ± 0.10 ^{Aa}	4.08 ± 0.08 ^{Aa}	4.10 ± 0.05 ^{Aa}	4.10 ± 0.10 ^{Aa}	4.13 ± 0.02 ^{Aa}
15	4.08 ± 0.08 ^{Aa}	4.03 ± 0.05 ^{Aa}	4.09 ± 0.01 ^{Aa}	4.06 ± 0.05 ^{ABa}	4.12 ± 0.00 ^{Aa}
30	4.05 ± 0.05 ^{ABa}	4.04 ± 0.05 ^{Aa}	4.09 ± 0.00 ^{Aa}	4.06 ± 0.05 ^{ABa}	4.11 ± 0.00 ^{Aa}
45	3.89 ± 0.09 ^{Bb}	3.70 ± 0.00 ^{Ba}	3.90 ± 0.10 ^{ABb}	4.03 ± 0.05 ^{ABc}	4.11 ± 0.01 ^{Ac}
60	3.60 ± 0.10 ^{Cab}	3.50 ± 0.10 ^{Ca}	3.83 ± 0.28 ^{BCbc}	4.01 ± 0.00 ^{Ac}	4.00 ± 0.00 ^{Bc}
75	3.50 ± 0.10 ^{CDa}	3.46 ± 0.05 ^{Ca}	3.63 ± 0.15 ^{Ca}	3.97 ± 0.06 ^{Bb}	3.90 ± 0.10 ^{Cb}
90	3.36 ± 0.06 ^{Da}	3.45 ± 0.05 ^{Ca}	3.86 ± 0.05 ^{ABCb}	3.97 ± 0.06 ^{Bb}	3.90 ± 0.10 ^{Cb}

The results are given as means ± SD, n = 3. a, b, c: In each row, values with the same letter (lower case) are not significantly different ($p > 0.05$). A, B, C, D: In each column, values with the same letter (upper case) are not significantly different ($p > 0.05$).

3.3. Color Analysis

The collected data of color parameters including L* (lightness), b* (yellowness), and a* (redness) during the storage period are presented in Figure 1. On day 0, all samples showed different values for color parameters. Lightness is the most important color parameter in sauce products; as it directly affects consumer choices [26]. Results showed that the L* factor had the highest value in the control sample (100SO) indicating that it was the most transparent sample among the others. This value decreased as the relative ratio of olive oil increased, such that samples with 75% and 100% olive oil had major differences from the other samples. The differences in the L* value is due to the different light scattering of the components from different samples.

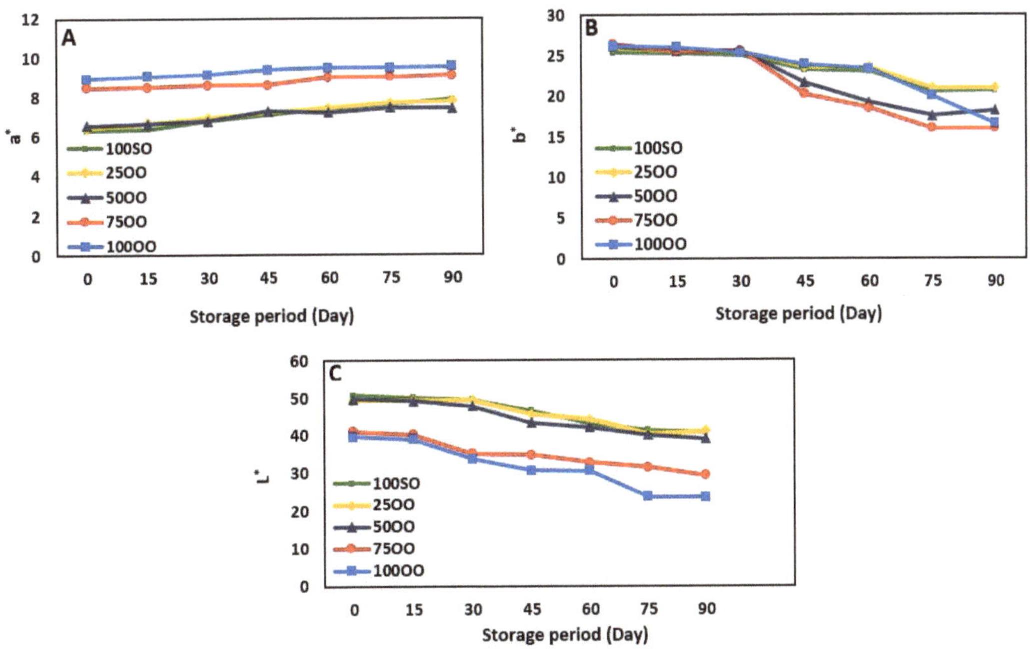

Figure 1. Color parameter ((**A**): a*: redness, (**B**): b*: yellowness, and (**C**): L*: lightness) changes in French sauce samples prepared with different blends of olive and soybean oil (100SO: samples with 100% soybean oil, 25OO: samples with 75% soybean oil and 25% olive oil, 50OO: samples with 50% soybean oil and 50% olive oil, 75OO: samples with 24% soybean oil and 75% olive oil, and 100OO: samples with 100% olive oil).

The scattering and absorption of samples are largely dependent on the reflective index, size and dispersion of droplets, concentration, and the presence of chromophobic materials [35]. As the soybean oil used in the current study was much paler than the olive oil (since soybean oil had a much lower content of pigments than did the olive oil, Table 1), the control sample showed maximum transparency (L* = 51). However, the sample 100OO showed the highest turbidity (L* = 40). Regarding the yellowness and redness factors, it can be clearly seen that the values increased with the increasing ratio of olive oil. The control sample showed the lowest values for b* and a* (25 and 6, respectively). In terms of b* value, there were no apparent differences among all samples on day 0 of production. The differences in the a* values, especially in the samples 75OO and 100OO, are due to the high amounts of intrinsic pigments in the olive oil [36]. On day 90 of storage, the lightness and yellowness values for all the samples were lowest, while the redness was somewhat higher. The rates of decrease in the samples containing higher levels of olive oil were higher if compared to that of the control sample. The polyphenolic content in olive oil is much higher than that in soybean oil (178 mg compared to 4 mg gallic acid equivalents/kg oil (Table 1)). As a result of the oxidation of polyphenolic compounds, the L* value can decrease faster in samples containing higher levels of olive oil. Additionally, the changes in the droplet sizes for the samples with different ratios of olive/soybean oils could be another reason for such changes. Generally, droplets get larger as the time proceeds towards the end of the storage period and therefore influences the light scattering of the samples [27]. This explanation was also confirmed by our emulsion stability results (Table 2), where a slight reduction was observed in the emulsion stability during the storage time. The b* values of all samples decreased on day 90 as a result of oxidation in the pigments [37]. Carotenoids (especially xanthophylls, which can be found in egg yolk and oils) are more sensitive to oxidation [38]. As a consequence, in samples containing higher levels of olive oil the degradation was much sharper than in the control. Badr [39] reported that the carotenoid content tends to decrease in liquid egg yolk during storage, but the rate of decrease at low temperatures is noticeably faster than at room temperature. The increase in the a* value during the storage time could be related to the oxidation of lipid components [27].

3.4. Textural Properties

The data obtained for the changes in the consistency and firmness of FS samples over the 90 days of storage are given in Table 4. On day 0, a decreasing trend was observed with an increase in the ratio of olive oil, which might be due to the lower amounts of unsaturated fatty acids in the samples containing higher levels of olive oil. Patil et al. [27] and Kupongsak et al. [40] reported similar results for the textural properties of mayonnaise at different ratios of fish/coconut and rice bran/coconut oil, respectively.

Table 4. Textural properties of French sauce samples prepared with different blends of olive and soybean oil (100SO: samples with 100% soybean oil, 25OO: samples with 75% soybean oil and 25% olive oil, 50OO: samples with 50% soybean oil and 50% olive oil, 75OO: samples with 25% soybean oil and 75% olive oil, and 100OO: samples with 100% olive oil).

Sample	Firmness (g)			Consistency (g.s)		
	Day 0	Day 45	Day 90	Day 0	Day 45	Day 90
100SO	163 ± 9 Aa	154 ± 3 Aa	128 ± 3 ABb	2507 ± 22 Aa	2303 ± 16 Ab	1938 ± 16 Ac
25OO	154 ± 5 ABa	146 ± 7 Aa	127 ± 3 ABb	2474 ± 15 Aa	2284 ± 18 Ab	1921 ± 22 Ac
50OO	161 ± 5 Aa	152 ± 2 Aa	131 ± 8 Ab	2407 ± 12 Ba	2180 ± 21 Bb	1952 ± 28 Ac
75OO	145 ± 4 BCa	130 ± 2 Bb	121 ± 9 ABb	2034 ± 27 Ca	1952 ± 17 Cb	1509 ± 11 Bc
100OO	134 ± 7 Ca	121 ± 4 Cb	116 ± 3 Bb	1829 ± 14 Da	1522 ± 11 Db	1488 ± 17 Bc

The results are given as means ± SD, $n = 3$. a, b, c: In each row, values with the same letter (lower case) are not significantly different ($p > 0.05$). A, B, C, D: In each column, values with the same letter (upper case) are not significantly different ($p > 0.05$).

During the study period (90 days), both the firmness and consistency dropped for all FS samples (Table 4). This fact could be related to a phenomenon such as coalescence, where oil droplets tend to join together and build larger particles [41]. These findings agree well with the results of the emulsion stability for the FS (Table 2).

3.5. Rheological Properties

The changes in the viscoelastic properties of FS samples are shown in Figure 2 (Storage modulus) and Figure 3 (Loss modulus). For all samples, the loss modulus (G") was lower than the storage modulus (G'). This indicates that all samples have an elastic behavior rather than a viscous behavior. Moreover, samples showed linear viscoelastic responses with increases in the angular frequency, indicating that all prepared FSs can be considered as a gel-like network [42].

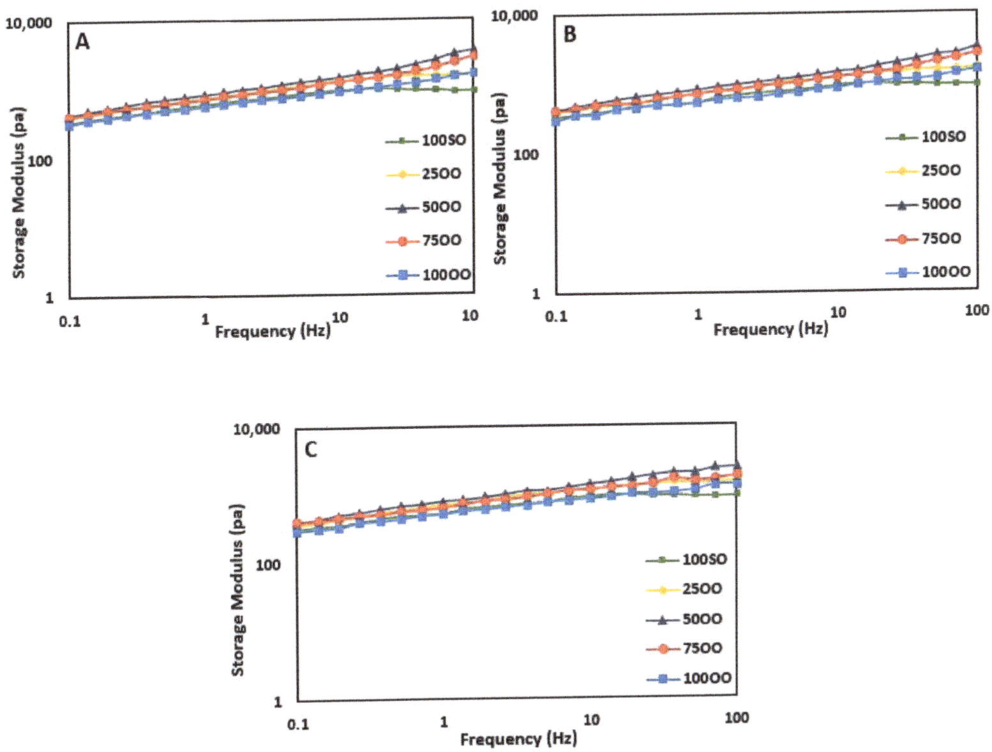

Figure 2. Storage modulus (G') of French sauce samples prepared with different blends of olive/soybean oil ((**A**): Day 0, (**B**): Day 45, and (**C**): Day 90 of storage). 100SO: samples with 100% soybean oil, 25OO: samples with 75% soybean oil and 25% olive oil, 50OO: samples with 50% soybean oil and 50% olive oil, 75OO: samples with 25% soybean oil and 75% olive oil, and 100OO: samples with 100% olive oil.

On day 0, the storage moduli for all samples increased slightly with an increase in the angular frequency (Figure 2). The sample 50OO and the control (100SO) indicated the highest and lowest storage modulus, respectively. This shows that the sample with 50% olive oil had the best power of structural uniformity [38], meaning that it required higher shear stress to flow. This subject is a matter of importance in the pouring ability of sauces in factories as well as packages.

In terms of loss modulus (Figure 3), there were no clear differences among the samples. However, sample 50OO showed the highest value. Higher values of loss modulus indicate that the sample containing 50% olive oil had a better behavior of viscoelasticity compared to the others. On day 45 and day 90, all samples showed a slight decrease in the storage modulus compared to day 0. This means that the elastic behavior and the structural uniformity dropped in all samples and samples had more liquid-like behaviors compared to the freshly-prepared samples on day 0. These findings are in agreement with the emulsion stability data, where the stability was reduced as the bonds between the oil droplets started to weaken. No apparent changes were found in terms of loss modulus during the storage time (Figure 3a–c).

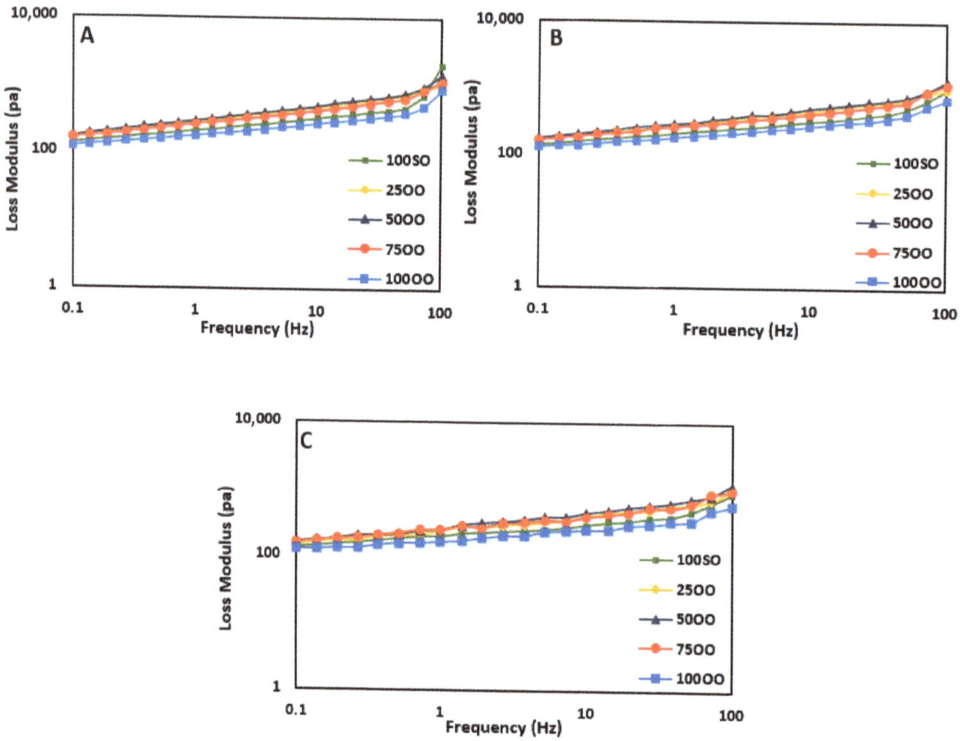

Figure 3. Loss modulus (G″) of French sauce samples prepared with different blends of olive/soybean oil ((**A**): Day 0, (**B**): Day 45, and (**C**): day 90 of storage) (100SO: samples with 100% soybean oil, 25OO: samples with 75% soybean oil and 25% olive oil, 50OO: samples with 50% soybean oil and 50% olive oil, 75OO: samples with 25% soybean oil and 75% olive oil, and 100OO: samples with 100% olive oil).

The flow curves of prepared FS samples with different ratios of olive/soybean oil are presented in Figure 4. The shear stress of all samples soared as the shear rate increased. However, there was no linearity in the relationship between the shear rate and shear stress for any of the sauce samples. Thus, all sauce samples can be considered as non-Newtonian liquids [43].

On day 0, sample 50OO showed the highest shear stress. This shows that it had the highest viscosity compared to the other samples; since the viscosity can be defined as the ratio of shear stress and shear rate. On days 45 and 90, all samples showed lower shear stress values compared to the freshly prepared samples on day 0. Generally, based on consumer data, higher values of shear stress as a function of shear rate are preferred [44]. Such products could cause a better mouth feel, but they could have some limitations considering the pumps and the fillers that are needed industrially [27].

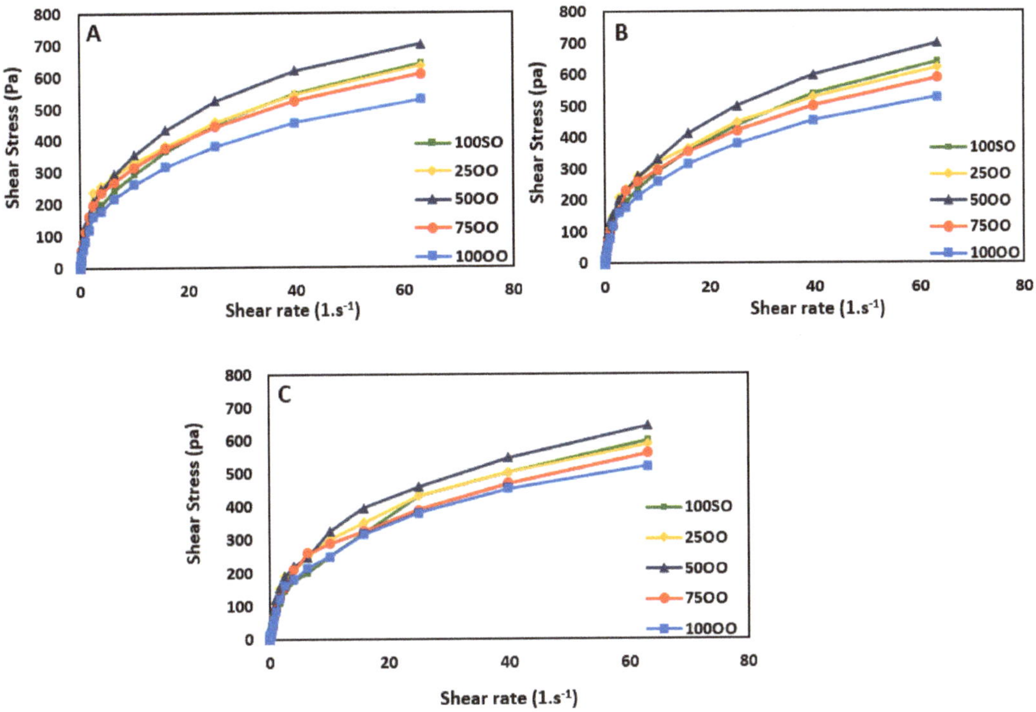

Figure 4. Flow curves of French sauce samples prepared with different blends of olive/soybean oil (**A**): Day 0, (**B**): Day 45, and (**C**): Day 90 of storage (100SO: samples with 100% soybean oil, 25OO: samples with 75% soybean oil and 25% olive oil, 50OO: samples with 50% soybean oil and 50% olive oil, 75OO: samples with 25% soybean oil and 75% olive oil, and 100OO: samples with 100% olive oil).

3.6. Stability Evaluation

3.6.1. Oxidative Parameters

The AVs of extracted lipid phases from the samples are shown in Table 5. Generally, AV measures the free fatty acids, which are made through the hydrolysis of lipids in the presence of water [45]. According to Table 5, the AV was increased in all samples during storage for 90 days, but the rates of increase were different among the different samples. The control sample (with no added OLE) showed the highest changes during storage, for which the AV was 1.1 at the beginning, but reached its maximum (4.9) at the end of the studied period. The increase in the AV during storage could be related to the activity of the hydrolytic enzymes in the eggs due to the acidic conditions of the sauce samples [46]. Additionally, the growth and activity of acid tolerant bacteria, especially lactic acid bacteria, could also impact on the free fatty acid formation in the aqueous phase of the samples [34]. The AV for the samples containing OLE, especially those with higher levels (1500 and 2000 mg/kg), increased at much lower rates than the control. Such effects can probably be mainly attributed to the polyphenolic compounds in the OLE, which were measured at 27 mg GAE/g dry leaf.

Table 5. Acid values (mg NaOH/g) obtained for the samples of French sauce (5OOO) enriched by different amounts of olive leaf extract (OLE).

Storage Period (Day)	OLE Concentration (mg/kg)				
	0	500	1000	1500	2000
0	1.1 ± 0.0 Aa	1.1 ± 0.0 Aa	1.1 ± 0.0 Aa	1.1 ± 0.0 Aa	1.1 ± 0.0 Aa
15	1.2 ± 0.0 Bab	1.2 ± 0.0 Ab	1.2 ± 0.01 Aab	1.1 ± 0.1 Aa	1.1 ± 0.0 Aa
30	1.4 ± 0.0 Ca	1.4 ± 0.0 Ba	1.3 ± 0.0 Aa	1.3 ± 0.0 Ba	1.3 ± 0.0 Ba
45	2.3 ± 0.0 Db	2.3 ± 0.1 Cab	2.3 ± 0.0 Bab	2.3 ± 0.0 Cb	2.2 ± 0.0 Ca
60	3.1 ± 0.0 Ec	3.1 ± 0.1 Dc	3.1 ± 0.0 Cc	2.9 ± 0.0 Db	2.6 ± 0.1 Da
75	3.7 ± 0.1 Fb	3.6 ± 0.1 Eb	3.6 ± 0.4 Db	3.0 ± 0.0 Da	2.9 ± 0.0 Ea
90	4.9 ± 0.0 Gd	5.0 ± 0.0 Fd	4.7 ± 0.1 Ec	3.7 ± 0.2 Eb	3.1 ± 0.1 Fa

The results are given as means ± SD, $n = 3$. a, b, c, d: In each row, values with the same letter (lower case) are not significantly different ($p > 0.05$). A, B, C, D, E, F, G: In each column, values with the same letter (upper case) are not significantly different ($p > 0.05$).

The results for changes in the PV during the storage period are shown in Table 6. The PV for all of the samples showed an initial increase followed by a slight decrease (except for the sample with 2000 mg/kg OLE) and then another increase afterward. The slight decrease, especially around day 30 and 45 of storage, is due to the breakdown of hydroperoxides that can form other compounds, and the high increase in the rate in samples around day 75 and 90 could be the result of boosting oxidative reactions [47]. Samples with higher amounts of OLE showed lower levels of PV than the control sample. While the sample containing 2000 mg/kg of OLE indicated the lowest PV value of 5.0 on day 90, the control sample showed the highest PV (8.9). These findings confirm the positive effect of OLE, especially its high and unique polyphenolic profile [48], on the sample's shelf life. The obtained results are in agreement with those of other related studies in this area [49,50].

Table 6. Peroxide values (meq O_2/kg) of French sauce (5OOO) samples enriched by olive leaf extract (OLE).

Storage Period (Day)	OLE Concentration (mg/kg)				
	0	500	1000	1500	2000
0	3.1 ± 0.0 Aa	3.1 ± 0.0 Aa	3.1 ± 0.0 Aa	3.1 ± 0.0 Aa	3.1 ± 0.0 Aa
15	3.8 ± 0.0 Ba	3.8 ± 0.0 Ba	3.7 ± 0.0 BCb	3.4 ± 0.0 Bc	3.2 ± 0.1 Bd
30	4.4 ± 0.0 Ca	4.3 ± 0.1 Ca	3.9 ± 0.0 Cb	3.6 ± 0.1 Dc	3.3 ± 0.0 Bd
45	3.9 ± 0.1 Bab	3.9 ± 0.1 Ba	3.8 ± 0.0 Bb	3.5 ± 0.1 Cc	3.3 ± 0.0 Bd
60	4.9 ± 0.1 Da	4.7 ± 0.1 Db	4.5 ± 0.1 Dc	4.2 ± 0.1 Ed	3.8 ± 0.0 Ce
75	6.1 ± 0.1 Ea	6.1 ± 0.0 Ea	5.9 ± 0.1 Eb	5.1 ± 0.1 Fc	4.5 ± 0.1 Dd
90	8.9 ± 0.0 Fa	8.3 ± 0.2 Fb	6.1 ± 0.1 Fc	4.9 ± 0.0 Gd	5.0 ± 0.1 Ed

The results are given as means ± SD, $n = 3$. a, b, c, d, e: In each row, values with the same letter (lower case) are not significantly different ($p > 0.05$). A, B, C, D, E, F, G: In each column, values with the same letter (upper case) are not significantly different ($p > 0.05$).

Generally, the oxidation process has a detrimental impact on products with high fat levels such as sauces. The reactions taking place in the interface of oil–water phases can speed up the process of oxidation [51].

PAV is another parameter that measures the secondary oxidative compounds, which are responsible for the undesirable flavors and odor in products. These oxidation reactions can lead to the formation of numerous compounds, especially 2-alkenals and 2,4-dienals [50]. Generally, high levels of PAV indicate the deterioration of lipids in the food matrix [52].

The PAV results of samples during the storage period are presented in Table 7. There was an increasing trend in the PAV during the storage time of 90 days. Such a trend could be related to the acceleration of oxidative processes with time. Samples containing higher levels of OLE (1500 and 2000 mg/kg) showed lower increases in the PAV. Timm-Heinrich et al. [53] reported that some polyphenolic compounds can minimize the production of oxidative products such as hexanal and heptadienal. On day 90 of storage in the current study, the control sample had the highest PAV, while the sample containing 2000 mg/kg of OLE showed the lowest PAV (12.6 and 5.7, respectively).

Table 7. *p*-Anisidine values (mmol/kg) of French sauce samples (5000) enriched by different amounts of olive leaf extract (OLE).

Storage Period (Day)	OLE Concentration (mg/kg)				
	0	500	1000	1500	2000
0	3.3 ± 0.0 ^{Aa}	3.3 ± 0.01 ^{Aa}	3.3 ± 0.0 ^{Aa}	3.3 ± 0.0 ^{Aa}	3.3 ± 0.0 ^{Aa}
15	5.9 ± 0.1 ^{Bc}	5.8 ± 0.0 ^{Bc}	5.84 ± 0.1 ^{Bc}	4.0 ± 0.1 ^{ABb}	3.8 ± 0.1 ^{Ba}
30	8.0 ± 0.0 ^{Cb}	7.9 ± 0.1 ^{Cb}	7.9 ± 0.8 ^{Cb}	3.9 ± 0.9 ^{ABa}	3.84 ± 0.1 ^{Ba}
45	8.1 ± 0.0 ^{Cb}	8.0 ± 0.0 ^{Cb}	8.0 ± 0.0 ^{Cb}	4.4 ± 0.1 ^{Ba}	4.2 ± 0.3 ^{Ca}
60	10.9 ± 0.0 ^{Dd}	10.8 ± 0.0 ^{Dd}	9.8 ± 0.6 ^{Db}	5.6 ± 0.6 ^{Ca}	5.0 ± 0.0 ^{Da}
75	11.3 ± 0.0 ^{Ed}	11.2 ± 0.0 ^{Ed}	10.0 ± 0.0 ^{Dc}	6.0 ± 0.0 ^{CDb}	5.1 ± 0.1 ^{Da}
90	12.6 ± 0.0 ^{Fd}	12.6 ± 0.0 ^{Fd}	10.9 ± 0.5 ^{Ec}	6.6 ± 0.7 ^{Db}	5.7 ± 0.2 ^{Ea}

The results are given as means ± SD, $n = 3$. a, b, c, d: In each row, values with the same letter (lower case) are not significantly different ($p > 0.05$). A, B, C, D, E, F: In each column, values with the same letter (upper case) are not significantly different ($p > 0.05$).

The total oxidation value (totox) is an index that is directly linked to both PV and PAV. Therefore, hydroperoxides and their primary and secondary oxidation products are reported by this parameter. The results of changes in the totox values of samples during the storage period are depicted in Figure 5. The totox values for all samples were increased by time. However, the samples with higher amounts of OLE (1500 and 2000 mg/kg) increased at lower rates (Figure 5). However, the control sample and sample with 500 mg/kg OLE experienced higher changes in terms of totox values. On day 90 of storage, the control sample indicated the highest totox value of 30.6 while the sample containing 2000 mg/kg OLE showed the lowest totox value of 15.8. Overall, the trend for changes in the totox value (Figure 5) is similar to that of PV and PAV (Tables 6 and 7, respectively). The decreasing effect of OLE on totox values in the samples treated with high amounts of OLE can be related to the presence of polyphenolic groups in the extract that have inhibitory effects against the generation of primary and secondary oxidation products [50].

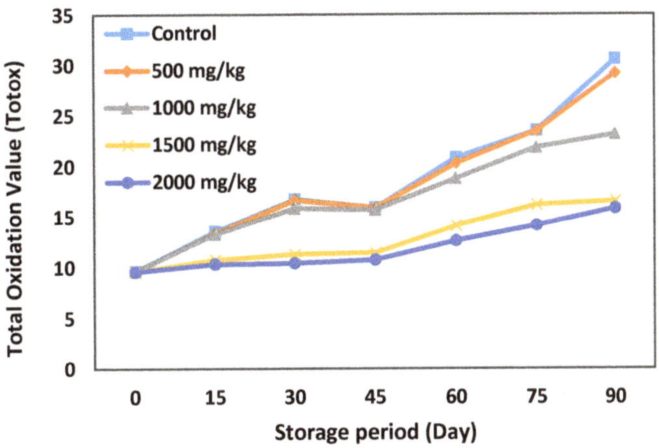

Figure 5. Changes in the totox value of French sauce samples enriched by different levels of olive leaf extract (OLE) (up to 2000 mg/kg) during 90 days of storage.

3.6.2. Microbiological Properties

The results of the total plate count for mesophilic aerobic bacteria of FS samples with and without using OLE are depicted in Figure 6. Both samples showed increases in the total plate counts during storage. However, the sample containing OLE at 2000 mg/kg indicated a lower value than the control. On day 0, the total plate count of the sample FS (i.e., that with no OLE added) was higher (300 cfu/g) than that in sample FS-OLE (containing OLE at 2000 mg/kg) (250 cfu/g). However, around day 15 of the storage, the total plate count for sample FS surpassed that of sample FS-OLE. Slight reductions were observed in the total counts of both samples around day 30 and 60 of storage, which could be due to the dissociation of acetic acid in the oil phase resulting in the depletion of microorganisms for a while [34]. The inhibitory effects of OLE on the growth of aerobic microorganisms can be due to its high levels of polyphenolic compounds that have been shown to have antibacterial and preservative effects in other studies [19,54]. Pourkomailian [55] reported that in samples with a high amount of oil phase, microorganisms usually stay in the oil phase and therefore they are not influenced by the pH effect of the aqueous phase.

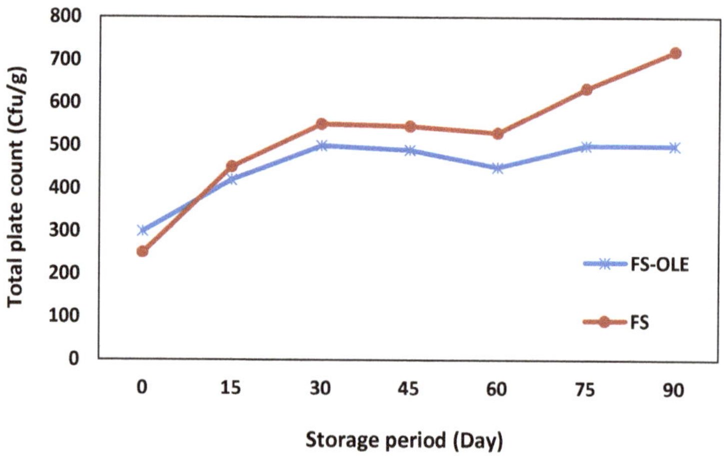

Figure 6. Total plate counts of mesophilic aerobic bacteria of sauce samples (FS-OLE: samples with 2000 mg/kg OLE and no commercial preservatives; FS: Samples with commercial preservatives, sodium benzoate, 0.063% (w/w), and potassium sorbate, 0.012% (w/w), without the addition of OLE).

A set of comparative microbiological data are shown in Table 8 for a sample containing 2000 mg/kg of OLE and no commercial preservatives, and a sample containing sodium benzoate at 0.063%, w/w, and potassium sorbate at 0.012%, w/w, as commercial synthetic preservatives, without the addition of OLE. According to these data, *E. coli* was not detected in either sample during the entire storage period. Moreover, the number of yeast and molds were similar in both samples (Table 8). Regarding acid tolerant bacteria, specifically lactic acid bacteria, they grew after day 60 of storage in sample FS; while in sample FS-OLE, they were not observed during the whole study period. These findings agree with the AV results of samples in Section 3.6 and confirms the protective role of olive leaves' polyphenolic content against acid tolerant bacteria, which could have undesirable effects on the samples.

Table 8. Total plate counts (cfu/g) of sauce samples during 90 days of storage.

	FS-OLE			FS		
	Yeast and Molds	*E. coli*	Lactic Acid Bacteria	Yeast and Molds	*E. coli*	Lactic Acid Bacteria
0	<100	ND	ND	<100	ND	ND
15	<100	ND	ND	<100	ND	ND
30	<100	ND	ND	<100	ND	ND
45	<100	ND	ND	<100	ND	ND
60	<100	ND	ND	<100	ND	4.0×10^2
75	<100	ND	ND	<100	ND	4.2×10^2
90	<100	ND	ND	<100	ND	4.2×10^2

3.7. Sensory Evaluation

The results for the sensory evaluation of sauce samples with and without enrichment by OLE at 2000 mg/kg and a commercial FS are presented in Figure 7. No significant differences were found among the samples in terms of their appearances, colors, and textures. Apparently, this level of OLE does not significantly influence these sensorial attributes. Additionally, the use of olive oil at 50% did not change these parameters either. The sample without OLE showed the lowest score in odor, but there was no major difference between the samples. However, the commercial product gained the best score for odor. With regard to the taste scores, the commercial product obtained the best score, while the sample containing OLE had the lowest score, which was due to the intrinsic taste of OLE which could even be perceived at low concentrations. Regarding the overall acceptance scores, the commercial product received the highest average score of 7.2 but the sample without OLE resulted in a score of 6.4, which was significantly different from the sample containing OLE receiving a score of 5.7.

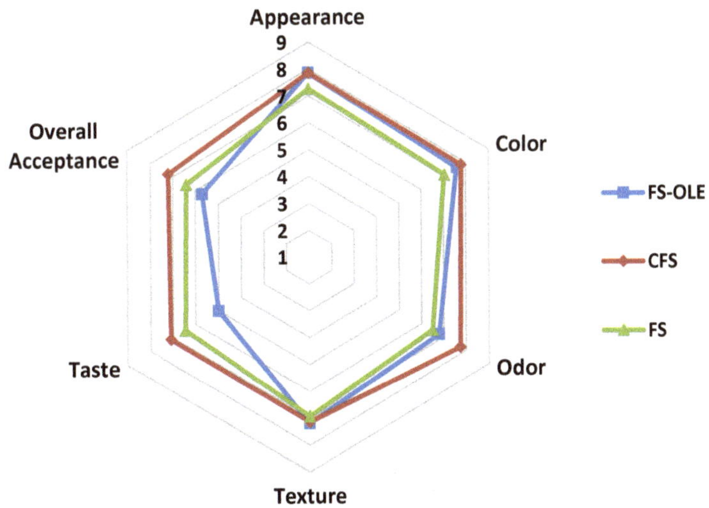

Figure 7. Sensory evaluation of French sauce samples (FS-OLE: samples with OLE, CFS: commercial French sauce, and FS: samples without OLE). Results are illustrated on a 9-point hedonic scale.

4. Conclusions

The major objective of this study was to produce a functional product replacing part of the soybean oil, which is currently used in the formulation of such products, with olive oil. Additionally, to improve the health aspects of the sauce, the enrichment of OLE in the sauce samples was investigated. Based on the results, there was a slight reduction in the emulsion stability of samples with increasing the amount of olive oil, especially at higher levels (75% and 100%). Moreover, the type and amount of the oils also affected the color parameters due to the different amounts of pigment available in the two studied oils. Textural and viscoelastic properties of samples were affected by the addition of olive oil. It was found that the sample prepared with 50% olive oil and 50% soybean oil was the best sample. Regarding the microbiological tests, it was understood that OLE was able to retard the growth of microorganisms during the storage time as well as prevent the growth of lactic acid bacteria at the end of the studied shelf-life time. So, OLE was found to be a promising alternative to commercial synthetic additives that are usually used in the formulation of sauces. Sensory analysis showed that OLE can influence the taste and odor even in small amounts, so further studies are needed to mask the intrinsic taste and odor of OLE in commercial food products.

Author Contributions: M.A.A. was the graduate student who carried out the experiments and prepared the draft of the manuscript under the supervision of the principal investigator, K.R., who was also in charge of the entire project. All authors have read and agreed to the published version of the manuscript.

Funding: The financial support provided by the research council of the College of Agriculture and Natural Resources at the University of Tehran is gratefully appreciated.

Institutional Review Board Statement: Not applicable.

Informed Consent Statement: Not applicable.

Conflicts of Interest: There are no conflict to declare.

References

1. Mizani, M.; Yaghoti Moghaddam, M.; Alimi, M.; Salehifar, M. Particle Size Distribution and Viscoelastic Behavior of French Dressing Containing Two Types of Commercial Waxy Maize Starches. *J. Food Biosci. Technol.* **2015**, *5*, 1–10.
2. Mcclements, D.J. *Food Emulsions: Principles, Practices, and Techniques*, 3rd ed.; CRC Press: New York, NY, USA, 2015; ISBN 9781498726696.
3. Rasmussen, C.N.; Wang, X.-H.; Leung, S.; Andrae-Nightingale, L.M.; Schmidt, S.J.; Engeseth, N.J. Selection and Use of Honey as an Antioxidant in a French Salad Dressing System. *J. Agric. Food Chem.* **2008**, *56*, 8650–8657. [CrossRef]
4. de Melo, A.N.F.; de Souza, E.L.; da Silva Araujo, V.B.; Magnani, M. Stability, nutritional and sensory characteristics of French salad dressing made with mannoprotein from spent brewer's yeast. *LWT-Food Sci. Technol.* **2015**, *62*, 771–774. [CrossRef]
5. Rognså, G.H.; Rathe, M.; Paulsen, M.T.; Petersen, M.A.; Brüggemann, D.A.; Sivertsvik, M.; Risbo, J. Preparation methods influence gastronomical outcome of hollandaise sauce. *Int. J. Gastron. Food Sci.* **2014**, *2*, 32–45. [CrossRef]
6. Mirzanajafi-Zanjani, M.; Yousefi, M.; Ehsani, A. Challenges and approaches for production of a healthy and functional mayonnaise sauce. *Food Sci. Nutr.* **2019**, *7*, 2471–2484. [CrossRef]
7. Lenart, E.; Kiritsakis, A.; Willett, W. Nutritional and Health Aspects of olive oil and Diseases. In *Olives and Olive Oil as Functional Foods*; Shahidi, F., Apostolos, K., Eds.; John Wiley & Sons, Ltd.: Chichester, UK, 2017; pp. 483–504.
8. Battino, M.; Forbes-Hernández, T.Y.; Gasparrini, M.; Afrin, S.; Cianciosi, D.; Zhang, J.; Manna, P.P.; Reboredo-Rodríguez, P.; Varela Lopez, A.; Quiles, J.L.; et al. Relevance of functional foods in the Mediterranean diet: The role olive oil, berries and honey in the prevention of cancer and cardiovascular diseases. *Crit. Rev. Food Sci. Nutr.* **2019**, *59*, 893–920. [CrossRef]
9. Lopez-Huertas, E. Health effects of oleic acid and long chain omega-3 fatty acids (EPA and DHA) enriched milks. A review of intervention studies. *Pharmacol. Res.* **2010**, *61*, 200–207. [CrossRef]
10. Giacintucci, V.; Di Mattia, C.; Sacchetti, G.; Neri, L.; Pittia, P. Role of olive oil phenolics in physical properties and stability of mayonnaise-like emulsions. *Food Chem.* **2016**, *213*, 369–377. [CrossRef]
11. Abdulmumeen, H.A.; Risikat, A.N.; Sururah, A.R. Food: Its preservatives, additives and applications. *Int. J. Chem. Biochem. Sci.* **2012**, *1*, 36–47.
12. Rafiee, Z.; Barzegar, M.; Sahari, M.A.; Maherani, B. Nanoliposomes Containing Pistachio Green Hull's Phenolic Compounds as Natural Bio-Preservatives for Mayonnaise. *Eur. J. Lipid Sci. Technol.* **2018**, *120*, 1800086. [CrossRef]
13. Bruzewicz, S.; Malicki, A.; Oszmiański, J.; Jaroslawska, A.; Jarmoluk, A.; Pawlas, K. Baicalin, Added as the Only Preservative, Improves the Microbiological Quality of Homemade Mayonnaise. *Pak. J. Nutr.* **2005**, *5*, 30–33. [CrossRef]
14. Flamminii, F.; Di Mattia, C.D.; Difonzo, G.; Neri, L.; Faieta, M.; Caponio, F.; Pittia, P. From by-product to food ingredient: Evaluation of compositional and technological properties of olive-leaf phenolic extracts. *J. Sci. Food Agric.* **2019**, *99*, 6620–6627. [CrossRef]
15. Souilem, S.; El-Abbassi, A.; Kiai, H.; Hafidi, A.; Sayadi, S.; Galanakis, C.M. Olive oil production sector: Environmental effects and sustainability challenges. In *Olive Mill Waste*; Elsevier: Amsterdam, The Netherlands, 2017; Volume 21, pp. 1–28, ISBN 9780080449111.

16. Aliyari, M.A.; Salami, M.; Hosseini, E.; Emam-Djomeh, Z.; Karboune, S.; Waglay, A. Biophysical, Rheological, and Functional Properties of Complex of Sodium Caseinate and Olive Leaf Aqueous Polyphenolic Extract Obtained Using Ultrasound-Assisted Extraction. *Food Biophys.* **2021**, *16*, 325–336. [CrossRef]
17. Difonzo, G.; Russo, A.; Trani, A.; Paradiso, V.M.; Ranieri, M.; Pasqualone, A.; Summo, C.; Tamma, G.; Silletti, R.; Caponio, F. Green extracts from Coratina olive cultivar leaves: Antioxidant characterization and biological activity. *J. Funct. Foods* **2017**, *31*, 63–70. [CrossRef]
18. Sueishi, Y.; Nii, R. A comparative study of the antioxidant profiles of olive fruit and leaf extracts against five reactive oxygen species as measured with a multiple free-radical scavenging method. *J. Food Sci.* **2020**, *85*, 2737–2744. [CrossRef] [PubMed]
19. Thielmann, J.; Kohnen, S.; Hauser, C. Antimicrobial activity of Olea europaea Linné extracts and their applicability as natural food preservative agents. *Int. J. Food Microbiol.* **2017**, *251*, 48–66. [CrossRef] [PubMed]
20. Şahin, S.; Bilgin, M. Olive tree (*Olea europaea* L.) leaf as a waste by-product of table olive and olive oil industry: A review. *J. Sci. Food Agric.* **2018**, *98*, 1271–1279. [CrossRef]
21. Sikora, M.; Badrie, N.; Deisingh, A.K.; Kowalski, S. Sauces and Dressings: A Review of Properties and Applications. *Crit. Rev. Food Sci. Nutr.* **2008**, *48*, 50–77. [CrossRef] [PubMed]
22. AOCS. *Official Methods and Recommended Practices*, 5th ed.; Revisions and Corrections, 584 Official Method Ce 1h 05; AOCS Press: Champaign, IL, USA, 2005.
23. Sousa, A.; Casal, S.; Malheiro, R.; Lamas, H.; Bento, A.; Pereira, J.A. Aromatized olive oils: Influence of flavoring in quality, composition, stability, antioxidants, and antiradical potential. *LWT-Food Sci. Technol.* **2015**, *60*, 22–28. [CrossRef]
24. Issaoui, M.; Flamini, G.; Hajaij, M.E.; Cioni, P.L.; Hammami, M. Oxidative Evolution of Virgin and Flavored Olive Oils under Thermo-oxidation Processes. *J. Am. Oil Chem. Soc.* **2011**, *88*, 1339–1350. [CrossRef]
25. Şahin, S.; Şamlı, R. Optimization of olive leaf extract obtained by ultrasound-assisted extraction with response surface methodology. *Ultrason. Sonochem.* **2013**, *20*, 595–602. [CrossRef]
26. Mun, S.; Kim, Y.-L.; Kang, C.-G.; Park, K.-H.; Shim, J.-Y.; Kim, Y.-R. Development of reduced-fat mayonnaise using 4αGTase-modified rice starch and xanthan gum. *Int. J. Biol. Macromol.* **2009**, *44*, 400–407. [CrossRef]
27. Patil, U.; Benjakul, S. Physical and Textural Properties of Mayonnaise Prepared Using Virgin Coconut Oil/Fish Oil Blend. *Food Biophys.* **2019**, *14*, 260–268. [CrossRef]
28. Liu, H.; Xu, X.M.; Guo, S.D. Rheological, texture and sensory properties of low-fat mayonnaise with different fat mimetics. *LWT-Food Sci. Technol.* **2007**, *40*, 946–954. [CrossRef]
29. McClements, D.J.; Decer, E.A. Lipid Oxidation in Oil-in-Water Emulsions: Impact of Molecular Environment on Chemical Reactions in Heterogeneous Food Systems. *J. Food Sci.* **2000**, *65*, 1270–1282. [CrossRef]
30. Depree, J.; Savage, G. Physical and flavor stability of mayonnaise. *Trends Food Sci. Technol.* **2001**, *12*, 157–163. [CrossRef]
31. Magnusson, E.; Rosén, C.; Nilsson, L. Freeze–thaw stability of mayonnaise type oil-in-water emulsions. *Food Hydrocoll.* **2011**, *25*, 707–715. [CrossRef]
32. Ennouri, M.; Evelyne, B.; Laurence, M.; Hamadi, A. Fatty acid composition and rheological behaviour of prickly pear seed oils. *Food Chem.* **2005**, *93*, 431–437. [CrossRef]
33. Worrasinchai, S.; Suphantharika, M.; Pinjai, S.; Jamnong, P. β-Glucan prepared from spent brewer's yeast as a fat replacer in mayonnaise. *Food Hydrocoll.* **2006**, *20*, 68–78. [CrossRef]
34. Karas, R.; Skvarča, M.; Žlender, B. Sensory quality of standard and light mayonnaise during storage. *Food Technol. Biotechnol.* **2002**, *40*, 119–127.
35. McClements, D.J. Colloidal basis of emulsion color. *Curr. Opin. Colloid Interface Sci.* **2002**, *7*, 451–455. [CrossRef]
36. Karabagias, I.K.; Badeka, A.; Casiello, G.; Longobardi, F.; Kontominas, M.G. Rapid screening of olive oil cultivar differentiation based on selected physicochemical parameters, pigment content and fatty acid composition using advanced chemometrics. *Eur. Food Res. Technol.* **2019**, *245*, 2027–2038. [CrossRef]
37. Bechoff, A.; Westby, A.; Owori, C.; Menya, G.; Dhuique-Mayer, C.; Dufour, D.; Tomlins, K. Effect of drying and storage on the degradation of total carotenoids in orange-fleshed sweetpotato cultivars. *J. Sci. Food Agric.* **2010**, *90*, 622–629. [CrossRef]
38. Laca, A.; Sáenz, M.C.; Paredes, B.; Díaz, M. Rheological properties, stability and sensory evaluation of low-cholesterol mayonnaises prepared using egg yolk granules as emulsifying agent. *J. Food Eng.* **2010**, *97*, 243–252. [CrossRef]
39. Badr, H.M. Effect of gamma radiation and cold storage on chemical and organoleptic properties and microbiological status of liquid egg white and yolk. *Food Chem.* **2006**, *97*, 285–293. [CrossRef]
40. Kupongsak, S.; Sathitvorapojjana, S. Properties and Storage Stability of O/W Emulsion Replaced with Medium-Chain Fatty Acid Oil. *Pol. J. Food Nutr. Sci.* **2017**, *67*, 107–115. [CrossRef]
41. Dickinson, E. Structure and Rheology of Simulated Gels Formed from Aggregated Colloidal Particles. *J. Colloid Interface Sci.* **2000**, *225*, 2–15. [CrossRef]
42. Mancini, F.; Montanari, L.; Peressini, D.; Fantozzi, P. Influence of Alginate Concentration and Molecular Weight on Functional Properties of Mayonnaise. *LWT-Food Sci. Technol.* **2002**, *35*, 517–525. [CrossRef]
43. Nikzade, V.; Tehrani, M.M.; Saadatmand-Tarzjan, M. Optimization of low-cholesterol–low-fat mayonnaise formulation: Effect of using soy milk and some stabilizer by a mixture design approach. *Food Hydrocoll.* **2012**, *28*, 344–352. [CrossRef]
44. Trotsek, D. *Chemical Engineering for the Food Industry*; Fryer, P.J., Pyle, D.L., Rielly, C.D., Eds.; Food Engineering Series; Springer: Boston, MA, USA, 2012; ISBN 978-1-4613-6724-6.

45. Ghorbani Gorji, S.; Smyth, H.E.; Sharma, M.; Fitzgerald, M. Lipid oxidation in mayonnaise and the role of natural antioxidants: A review. *Trends Food Sci. Technol.* **2016**, *56*, 88–102. [CrossRef]
46. Honold, P.J.; Jacobsen, C.; Jónsdóttir, R.; Kristinsson, H.G.; Hermund, D.B. Potential seaweed-based food ingredients to inhibit lipid oxidation in fish-oil-enriched mayonnaise. *Eur. Food Res. Technol.* **2016**, *242*, 571–584. [CrossRef]
47. Sørensen, A.-D.M.; Nielsen, N.S.; Jacobsen, C. Oxidative stability of fish oil-enriched mayonnaise-based salads. *Eur. J. Lipid Sci. Technol.* **2010**, *112*, 476–487. [CrossRef]
48. Lockyer, S.; Rowland, I.; Spencer, J.P.E.; Yaqoob, P.; Stonehouse, W. Impact of phenolic-rich olive leaf extract on blood pressure, plasma lipids and inflammatory markers: A randomised controlled trial. *Eur. J. Nutr.* **2017**, *56*, 1421–1432. [CrossRef] [PubMed]
49. Alizadeh, L.; Abdolmaleki, K.; Nayebzadeh, K.; Shahin, R. Effects of tocopherol, rosemary essential oil and Ferulago angulata extract on oxidative stability of mayonnaise during its shelf life: A comparative study. *Food Chem.* **2019**, *285*, 46–52. [CrossRef]
50. Kishk, Y.F.M.; Elsheshetawy, H.E. Effect of ginger powder on the mayonnaise oxidative stability, rheological measurements, and sensory characteristics. *Ann. Agric. Sci.* **2013**, *58*, 213–220. [CrossRef]
51. Waraho, T.; McClements, D.J.; Decker, E.A. Mechanisms of lipid oxidation in food dispersions. *Trends Food Sci. Technol.* **2011**, *22*, 3–13. [CrossRef]
52. Park, B.I.; Kim, J.; Lee, K.; Lim, T.; Hwang, K.T. Flavonoids in common and tartary buckwheat hull extracts and antioxidant activity of the extracts against lipids in mayonnaise. *J. Food Sci. Technol.* **2019**, *56*, 2712–2720. [CrossRef]
53. Timm-Heinrich, M.; Eymard, S.; Baron, C.P.; Nielsen, H.H.; Jacobsen, C. Oxidative changes during ice storage of rainbow trout (Oncorhynchus mykiss) fed different ratios of marine and vegetable feed ingredients. *Food Chem.* **2013**, *136*, 1220–1230. [CrossRef] [PubMed]
54. Korukluoglu, M.; Sahan, Y.; Yigit, A. Antifungal properties of olive leaf extracts and their phenolic compounds. *J. Food Saf.* **2008**, *28*, 76–87. [CrossRef]
55. Pourkomailian, B. Sauces and dressings. In *The Stability and Shelf-Life of Food*; CRC Press: Washington, DC, USA, 2000; pp. 311–329, ISBN 0849308577.

life

MDPI

Article

Processing of Flavor-Enhanced Oils: Optimization and Validation of Multiple Headspace Solid-Phase Microextraction-Arrow to Quantify Pyrazines in the Oils

Ziyan Xu [1], Chuan Zhou [2], Haiming Shi [2], Hong Zhang [2], Yanlan Bi [1] and Xuebing Xu [2,*]

1 College of Food Science and Technology, Henan University of Technology, Lianhua Road, Zhengzhou 450001, China; zyxhghy@163.com (Z.X.); bylzry@126.com (Y.B.)
2 Wilmar (Shanghai) Biotechnology Research and Development Center Co., Ltd., 118 Gaodong Road, Pudong New District, Shanghai 200137, China; zhouchuan@cn.wilmar-intl.com (C.Z.); shihaiming@cn.wilmar-intl.com (H.S.); zhanghongsh@cn.wilmar-intl.com (H.Z.)
* Correspondence: xuxuebing@cn.wilmar-intl.com; Tel.: +86-21-3115-3256

Abstract: An efficient and effective multiple headspace-solid phase microextraction-arrow-gas chromatography-mass spectrometry (MHS-SPME-arrow-GCMS) analytical protocol is established and used to quantify the flavor compounds in oils. SPME conditions, such as fiber coating, pre-incubation temperature, extraction temperature, and time were studied. The feasibility was compared between SPME-arrow and the traditional fiber by loading different sample amounts. It was found that the SPME-arrow was more suitable for the MHS-SPME. The limit of detection (LODs) and limit of quantitation (LOQs) of pyrazines were in the range of 2–60 ng and 6–180 ng/g oil, respectively. The relative standard deviation (RSD) of both intra- and inter-day were lower than 16%. The mean recoveries for spiked pyrazines in rapeseed oil were in the range of 91.6–109.2%. Furthermore, this newly established method of MHS-SPME-arrow was compared with stable isotopes dilution analysis (SIDA) by using [$^2H_6$]-2-methyl-pyrazine. The results are comparable and indicate this method can be used for edible oil flavor analysis.

Keywords: solid-phase microextraction-arrow; multiple headspace solid-phase microextraction; pyrazine; flavor edible oil; internal standard method

Citation: Xu, Z.; Zhou, C.; Shi, H.; Zhang, H.; Bi, Y.; Xu, X. Processing of Flavor-Enhanced Oils: Optimization and Validation of Multiple Headspace Solid-Phase Microextraction-Arrow to Quantify Pyrazines in the Oils. *Life* **2021**, *11*, 390. https://doi.org/10.3390/life11050390

Academic Editor: Sabato D'Auria

Received: 7 April 2021
Accepted: 20 April 2021
Published: 26 April 2021

1. Introduction

Good natural flavor is one of the major drivers for consumers to purchase edible oils, especially in Asia. Flavor-enhanced edible oils are normally produced by high roasting temperature in mechanical pressing. Under high temperatures in the processing, a lot of flavor volatiles are created through various reactions, in particular the Mallard reactions. Pyrazines are a group of such volatiles that attract the most importance. A lot of studies have shown that pyrazines are formed by roasting, baking, or thermally processing through Maillard reactions and impart cocoa, peanut, roasted nut-like flavors to various foods [1–3].

Generally, pyrazines, which exhibit high odor activity values, are highly correlated with roasted flavor in roasted black tea, coffee, and oils [4–7]. For instance, the total pyrazine level was closely related to the roasted peanut flavor in peanut by sensory evaluation and correlation analysis [8]. Meanwhile, it was found that 2,5-dimethyl-pyrazine was one of the key flavor compounds for sesame oil (Odor Activity Value, OAV > 100) [6], while 2-ethyl-5(or 6)-methyl-pyrazine, 2,6-diethyl-pyrazine, and 3-ethyl-2,5-dimethyl-pyrazine were the major aroma contributors in the roasted sesame oil [9].

The above pyrazines were also found to be aroma-active compounds in flavor-enhanced rapeseed oils and roasted pumpkin seed oils [10,11]. Interestingly, almost the same group of pyrazines was found in various oils, and variation in the concentration of individual compounds made their flavor different from one to another. This is in line with Poehlmann's notion that impressions of different flavors of roasted oils are caused by differences in the quantity rather than the quality of their key aroma-active compounds [10].

The conventional methods of extracting the aroma compounds of oils mainly include headspace solid-phase microextraction (HS-SPME), solvent-assisted flavor evaporation (SAFE), simultaneous distillation extraction (SDE), dynamic headspace (DHS), static headspace (SHS), and stir bar sorptive extraction (SBSE) [12–15]. Compared to other methods, HS-SPME is a simple and easy technique, which reduces the size of extraction instrumentation and operates solvent-free [16], and has been widely used for volatiles analysis in food products, such as pasta [17], bread [18], and edible oils [19].

SPME-arrow fiber is a new device, which has 6–20 times larger than the sorbent phase of traditional SPME fiber and shows more than 10 times the sensitivities and extraction efficacies [20]. Until now, SPME-arrow fiber has been employed for the determination of volatiles of various food products, such as brown rice vinegar [21], Korean salt–fermented fish sauce [20], soy sauce [22], and milk [23]. In these studies, SPME-arrow showed higher extraction performance, such as more kinds of compounds extracted, larger extraction amount, and better reproducibility. Moreover, SPME-arrow was evaluated as more effective in extracting pyrazines in food matrixes.

The HS-SPME technique is based on the distribution of target analyte in the headspace, sample matrix, and SPME fiber coating. Compared to other methods, SPME is a non-exhaustive extraction technique, which means only a small portion of target analytes can be extracted every time [16]. Previous studies have shown that flavor-enhanced oils contain hundreds of volatile compounds [1,24,25]. For example, Kim has detected 80 peaks from the volatiles of roasted perilla seed oil [24], Liu has observed 118 peaks, and identified 94 compounds from the volatiles of roasted peanut oil [1]. In the sealed vials, the distribution of volatiles is different in the three phases, which results in the selectivity of different volatiles by HS-SPME. On the other hand, non-degummed and undeodorized flavor oil is a complex sample matrix. The phospholipids, free fatty acids, and other compositions in oils will affect the volatilization and distribution of volatiles by binding, which is called the "matrix effect" [26]. For example, phenolic compounds were found to have an effect on the release of volatiles in virgin olive oil [27], which means the composition and property of the sample will affect the distribution of volatiles in the three phases. When there is a matrix effect, HS-SPME combined with traditional calibration methods, such as external standard and internal standard, has some limitations in the quantitation of volatiles in flavor oils. The stable isotope dilution analysis (SIDA) is a good approach that can effectively overcome the matrix effect [28]. However, the stable isotope standards are expensive and not all available commercially.

Multiple headspace solid-phase microextraction (MHS-SPME), which was developed by multiple headspace extraction (MHE) and firstly proposed by Suzuki and McAuliffe, along with Kolb [29–31], is one of the applications of exhaustive extraction by HS-SPME. The principle of the MHS-SPME procedure is based on the consecutive extraction of the same sample several times, while the total theoretical peak areas of the analyte corresponding to exhaustive extraction can be estimated from each individual extraction. The "matrix effect" can be eliminated by "exhaustive extraction". Ezquerro et al. firstly applied MHS-SPME to quantify volatile organic compounds (VOCs) in multilayer food packaging materials in 2003 compared with the SA [32]. The results showed that MHS-SPME is a powerful method for the direct quantification of VOCs in solid samples. The comparison of the total peak areas of the calibration solution prepared with water and *n*-hexadecane showed that the peak area obtained by MHS-SPME were statistically equal, which proved that MHS-SPME could effectively eliminate the matrix effect [26]. Moreover, the use of MHS-SPME in wine flavor analysis proved that MHS-SPME could overcome the low repeatability when SPME was applied to the complex matrix [33]. To our knowledge, MHS-SPME combined with GC-MS has been successfully used to quantify volatile compounds in food products such as sausage, wine, herbal teas, roasted coffee, and so forth [33–35]. However, its application has not been performed in flavor-enhanced edible oils.

The aim of this study is to set up a novel and more accurate MHS-SPME-arrow-GC-MS method to quantify pyrazines in flavor-enhanced edible oils. This is the first time applying MHS-SPME-arrow to quantify flavor compounds in oils combined with internal standard (ISTD) while carried out in organic solvent for quantitative calibrations. There are three steps needed to develop the quantitative method based on MHS-SPME-arrow: (1) Optimizing the HS-SPME conditions for the maximum extraction efficiency and sensitivity improvement; (2) finding the appropriate sample loading suitable for MHS-SPME and clarifying the feasibility of SPME-arrow fiber in replacement of the traditional SPME fiber; and (3) validating the MHS-SPME-arrow method combined with the internal standard in solvent by selected conditions, in which the MHS-SPME-arrow and SIDA were compared using [$^2H_6$]-2-methyl-pyrazine.

2. Materials and Methods

2.1. Materials

Thirteen pyrazine standards (2-methyl-pyrazine, 2,5-dimethyl-pyrazine, 2,6-dimethyl-pyrazine, 2-ethyl-pyrazine, 2-ethyl-5-methyl-pyrazine, 2-ethyl-6-methyl-pyrazine, trimethyl-pyrazine, 2,6-diethyl-pyrazine, 3-ethyl-2,5-dimethyl-pyrazine, 2-ethyl-3,5-dimethyl-pyrazine, 2-methyl-3,5-diethyl-pyrazine, iso-propenyl-pyrazine, 5H-5-methyl-6,7-dihydrocyclopenta-pyrazine) and internal standard of 3-methyl-pyridine were purchased from Sigma-Aldrich Co. Ltd. (Shanghai, China). Isotopically labeled internal standard: [$^2H_6$]-2-methyl-pyrazine was purchased from CDN Isotopes Inc. (CDN, Canada). Caprylic capric triglyceride (ODO) was from Maitian Chemical Co. Ltd. (Shanghai, China).

Four SPME-arrow fibers, including polydimethylsiloxane (PDMS) (100 μm × 20 mm), polyacrylate (PA) (100 μm × 20 mm), polydimethylsiloxane/divinylbenzene (PDMS/DVB) (120 μm × 20 mm), polydimethylsiloxane/divinyl-benzene/carboxen (PDMS/DVB/CAR) (120 μm × 20 mm), and traditional SPME DVB/CAR/PDMS (50/30 μm × 10 mm) fibers were purchased from CTC Analytics AG (Zwingen, Switzerland). The holder of fibers was purchased from PAL System (Zwingen, Switzerland).

The refined pyrazine-free and flavor-enhanced rapeseed oils were obtained from the local supermarket, thus as the real samples of 3 flavor-enhanced oils, including peanut oil, sesame oil, and rapeseed oil. All oil samples were stored at 4 °C. Ethyl acetate was chromatographic grades.

2.2. Preparation of Standard Solutions

Standard solutions were prepared using ethyl acetate as the solvent. The combined stock solution of target analytes (13 pyrazines) and internal standard (3-methyl-pyridine) were prepared by weighing and storing in sealed amber glassware at 0 °C in the dark. The concentration of analytes (pyrazines and 3-methyl-pyridine) in stock solutions were approximately 1000 mg/L, and the standard calibration solution (approximately 25 mg/L) was prepared by diluting the stock solution accordingly.

2.3. Sample Preparation

A stock solution of the internal standard 3-methyl-pyridine (2000.0 mg/kg) was prepared by adding 20.0 mg of 3-methyl-pyridine to 10.0 g of ODO. Then, the stock solution of the internal standard was diluted to 50.0 mg/kg with ODO. The resulting solutions were stored at 0 °C in the dark.

The resulting internal standard solution (50.0 mg/kg) of 50.0 mg was added to 1.0 g of the oil sample, homogenized by vortex mixer afterward. The oil samples with internal standard (50.0 mg) were weighed into a 20 mL headspace vial sealed with a PTFE/silicone septum screw-cap. The vials were placed in the autosampler tray for HS-SPME analyses. The procedure for stable isotope dilution assay was the same as above.

2.4. MHS-SPME Conditions

For the HS-SPME method, the oil samples were pre-incubated at 80 °C for 20 min with the agitation speed of 450 rpm to release the volatile compounds prior to extraction, and then the fiber was exposed in the headspace of the vial at 50 °C for 50 min for equilibrium extraction. After the extraction of volatiles from the oil onto the fiber, the analytes were thermally desorbed from the fiber in the injector port of the chromatograph for 80 s and transferred to the chromatograph column where volatile compounds were separated. Finally, the analytes were identified and quantified by a mass spectrometer. After extraction and desorption, the SPME fiber was conditioned at 230 °C for 3 min. In the MHS-SPME method, the oil samples were taken 4 times at equal time intervals (of about 70 min).

2.5. GC–MS Conditions

A Combi PAL ingenious sample handling system (Ingenious Lab, Zwingen, Switzerland) used as an autosampler was mounted into the gas chromatograph. The change of liquid injector tool and SPME-arrow tool were the robotic steps for the analytical process. The gas chromatograph system was an Agilent 8890 coupled with a 5977B mass spectrometer (Agilent Technologies, Paolo Alto, CA, USA). A DB-FFAP analytical column (60 m × 0.25 mm, 0.25 μm) from Agilent Technologies was carried out to separate analytes. The SPME fiber device, which was used as an injector, was desorbed at high temperatures in the gas chromatography injector port to transfer analytes to gas chromatography device.

GC conditions: Fiber desorption time: 80 s; injector temperature: 230 °C; injection mode: Splitless; carrier gas: Helium (99.999% purity); flow rate: 1.0 mL/min. The oven temperature program was as follows: The initial temperature was 40 °C (1.5 min); the temperature was programmed from 40 to 100 °C (5 min) at 10 °C/min, from 100 to 150 °C (10 min) at 2 °C/min, from 150 to 185 °C at 5 °C/min, from 185 to 245 °C (8 min) at 20 °C/min, for a total running time of 65.5 min.

MSD conditions: MS was operated in EI mode (70 eV); acquisition was carried out in full scan and selected ion monitoring (SCAN&SIM) mode; and the selected ions were reported in Table 1. Other conditions include ion source temperature: 230 °C; quadrupole temperatures: 150 °C; and transfer line temperature: 280 °C. Data were collected and processed using MassHunter software (Agilent Technologies). Analytes were identified by retention time and selected ions, which are listed in Table 1.

2.6. Quantification of MHS-SPME

MHS-SPME is a process based on stepwise extraction of the analytes from the same sample, which can be seen as a combination of multiple headspace and headspace solid-phase microextraction [31,32]. After the same sample was consecutively extracted 4 times, the total peak area of analytes (A_T) can be calculated by Equation (1):

$$A_T = \sum_{i=1}^{i=1} A_i = \frac{A_1}{1-\beta} \tag{1}$$

where i is the number of extractions, A_i is the peak area of analytes in the ith extraction, A_1 is the peak area of analytes in the first extraction, β is a constant between zero and one ($0 \leq \beta < 1$), which can be calculated from Equation (2):

$$\ln A_i = (i-1)\ln\beta + \ln A_1 \tag{2}$$

There is a linear relationship between $\ln A_i$ and $i - 1$, where $\ln\beta$ is the slope of the linear plot and can be calculated from a limited number (3 or 4) of extractions. Thus, the total amount of analytes in the system can be quantified by ISTD.

2.7. Calculation of ISTD

The detector response factors were calculated by the ethyl acetate solution of pyrazines and the internal standard. A liquid injection mode was used in this process by GC-MS. The response factors were calculated before samples were analyzed by MHS-SPME.

2.8. Statistical Analysis

Results were analyzed using ANOVA carried out using SPSS Statistical Software 18.0 (SPSS, Chicago, IL, USA), and the confidence interval was taken as 95%. All figures were generated using Origin 9.0, Adobe Illustrator, and Adobe Photoshop.

3. Results and Discussion

3.1. Optimization of HS-SPME Conditions

The experiments were performed on the flavor-enhanced rapeseed oil. Variables such as type of SPME-arrow fiber, pre-incubation temperature, extraction temperature, and time were studied to optimize the performance of HS-SPME. The HS-SPME conditions were optimized on the basis of peak areas of analytes and performed three times repeatedly. Figure 1 shows that the influence of the type of the fiber coating, pre-incubation temperature, extraction temperature, and time on the total peak area of target pyrazines.

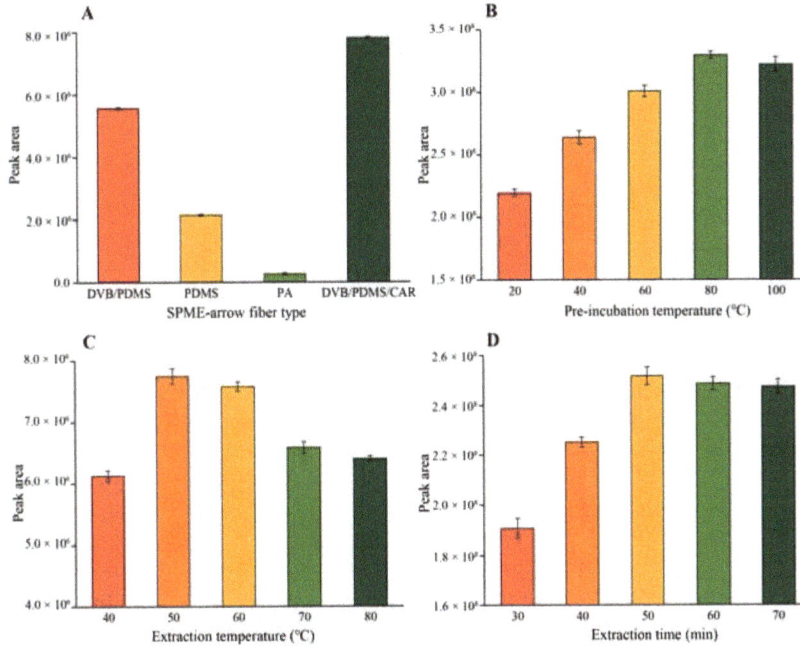

Figure 1. Influence of (**A**) the type of the fiber, (**B**) pre-incubation temperature, (**C**) extraction temperature, (**D**) extraction time on the total peak area of target compounds.

The sensitivity and selectivity of the extraction method were determined by the properties of adsorbents on SPME fibers. The selection of a suitable fiber was the first step in developing the HS-SPME method. The performance of SPME fibers for the target compounds depends on the polarity of the analytes and the physical and chemical properties of coating types [16]. In this study, the extraction efficiency of the four commercial SPME fibers with different polar coatings was compared under the same conditions. The results showed significant differences among the various fiber coatings ($p < 0.05$) (Figure 1A). PDMS/DVB/CAR, which is produced with a cross-linked coating and contains bipolar coatings thus that can be used for the extraction of polar and non-polar VOCs, had the best performance ($p < 0.05$). The single-phase of PDMS (non-polar) and PA (polar) showed the poorest performance. These results were consistent with previous studies [36]. Therefore, a PDMS/DVB/CAR (120 μm × 20 mm) fiber was chosen for the optimization of the HS-SPME conditions and further used in this study.

After the type of coating was selected, the second important parameter was pre-incubation temperature. The role of the pre-incubation procedure is to volatilize the aroma compounds from the sample and build equilibrium between the sample and headspace. At the same pre-incubation time, the increase of temperature can promote the distribution of weak volatile components in the headspace and accelerate the building of equilibrium. However, higher temperatures may lead to the conversion and degradation of unstable substances. Five pre-incubation temperatures were studied: 20, 40, 60, 80, and 100 °C. The results (Figure 1B) showed that the higher pre-incubation temperature indeed contributed to the volatilization and equilibrium of pyrazines. However, the total peak area was not significantly different between 80 °C and 100 °C. Thus, 80 °C for pre-incubation temperature was chosen as the SPME condition for further studies.

Extraction temperature is an important factor affecting the extraction efficiency of aroma compounds. The increasing temperature is conducive to the release of aroma compounds from the matrix at the same extraction time. However, high temperatures may also cause the degradation of components in the food matrix and decrease the absorbent ability of SPME fibers. Five extraction temperatures were studied: 40, 50, 60, 70, and 80 °C. According to Figure 1C, the extraction temperature of 50 °C was selected.

The effect of different extraction times (30, 40, 50, 60, and 70 min) was also evaluated at 50 °C. The optimum time was required to reach equilibrium in three phases: The fiber coating, the headspace, and the sample. SPME under equilibrium is an important condition for carrying out the MHS-SPME operation. Compared with extraction without equilibrium, extraction under equilibrium has better repeatability [32]. As shown in Figure 1D, 50 min was chosen as the optimal extraction time, which ensured extraction efficiency and established the equilibrium of pyrazines in three phases.

On that basis of the above experiments, the PDMS/DVB/CAR (120 μm × 20 mm) fiber coating was chosen, and the optimal HS-SPME conditions were pre-incubation temperature: 80 °C; extraction temperature: 50 °C; and extraction time: 50 min.

3.2. Amount of Oil Sample

The basic operation of MHS-SPME involves repeating extraction several times from the same sample, while the peak area decreases exponentially with the number of extractions. Therefore, the amounts of the sample should not only meet the LOQs but also ensure the linearity between $\ln A_i$ versus the extraction numbers $(i - 1)$, where $\ln A_i$ shows linear decay with the number of extractions (Equation (2)). In order to satisfy that requirement, $0.4 < \beta < 0.95$ should be fulfilled, which indicates the slope of the linear plot $\ln A_i$ versus $i-1$ must less than −0.0513 and more than −0.9163 [37]. Compared with the traditional SPME fiber, the SPME-arrow device has a larger sorbent phase and showed higher sensitivities and extraction amount [20]. In this study, the effects of seven sample amounts and two fiber types on β and coefficient of determination (R^2) were studied on flavor-enhanced rapeseed oils using the 20 mL HS vial.

For MHS-SPME, two conditions should be met: (1) $R^2 > 0.75$, (2) $0.4 < \beta < 0.95$, indicating that the $\ln A_i$ was linearly decayed with the number of extractions. Figure 2A shows the sample amount meeting $R^2 > 0.75$ while Figure 2B meeting $0.4 < \beta < 0.95$. As shown in Figure 2, traditional SPME fiber exhibited the most pyrazines that met requirements when the sample amount was 20.0 mg, while the SPME-arrow fiber showed the highest number at 20.0 mg and 50.0 mg samples. This was in line with the advantage of the SPME-arrow fiber that larger sorbent phase volume gives higher extraction capacity. When the sample amount was greater than 200 mg, almost no β satisfied the MHS-SPME applicable requirement, regardless if the traditional SPME fiber or SPME–arrow fiber was used. The reason might be that the headspace was saturated in the multiple headspace extractions. However, the low sample amount may lead to poor repeatability due to the mass of analytes closed to LOQs, which means that the SPME–arrow fiber was more suitable for the quantitation of pyrazines in flavor oils by MHS-SPME. Therefore, 50.0 mg was selected as the sample mass in the following study.

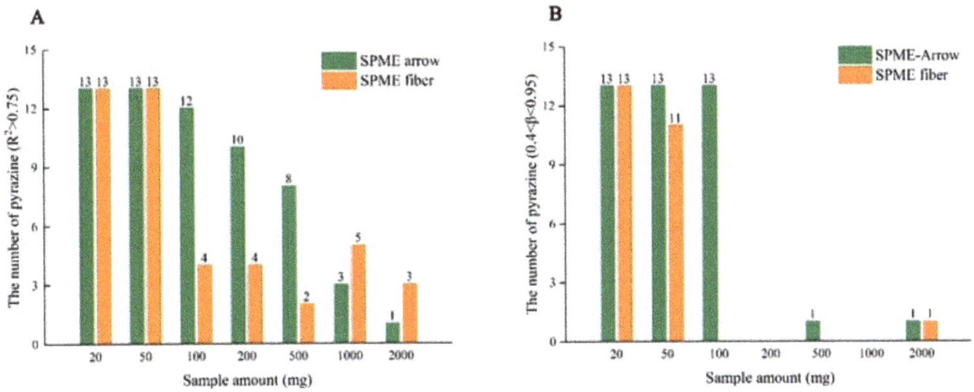

Figure 2. The number of pyrazines that met the following conditions: (**A**) $R^2 > 0.75$; (**B**) $0.4 < \beta < 0.95$, using solid phase microextraction (SPME)-arrow fiber and traditional SPME fiber, respectively.

3.3. Validation of MHS-SPME-Arrow Method

The analytical performance of the MHS-SPME-arrow-GC-MS was evaluated in terms of LOD, LOQ, inter-day precision, intra-day precision, and recovery, as shown in Tables 1 and 2.

Table 1. List of retention time, selected ions, limits of detection (LOD), limits of quantification (LOQ), inter-day precision, intra-day precision of 13 target pyrazines in oils.

Compound	Retention Time (min)	Selected Ions [1]	LOD [2] (ng/g)	LOQ [2] (ng/g)	Intra-Precision RSD (%)	Inter-Precision RSD (%)	Concentration Level [3] (ng/g)
3-methyl, pyridine [4]	15.93	66, 92, **93**	–	–	–	–	–
2-methyl-pyrazine	14.78	67, 53, **94**	3	6	7.13	8.66	907
2,5-dimethyl-pyrazine	16.62	42, 81, **108**	5	20	5.11	5.67	5011
2,6-dimethyl-pyrazine	16.90	40, 42, **108**	8	20	13.68	12.14	264
Ethyl-pyrazine	17.05	80, 108, **107**	2	6	12.28	14.77	192
2-ethyl-6-methyl-pyrazine	18.93	94,122, **121**	3	6	9.64	12.27	363
2-ethyl-5-methyl-pyrazine	19.21	94, 122, **121**	3	10	7.83	9.56	586
Trimethyl-pyrazine	19.80	42, 81, **122**	3	6	10.61	11.01	622
2,6-diethyl-pyrazine	21.04	108, 136, **135**	3	10	13.69	15.10	43
3-ethyl-2,5-dimethyl-pyrazine	21.41	108, 136, **135**	4	20	11.28	10.78	1837
2-ethyl-3,5-dimethyl-pyrazine	22.28	108, 136, **135**	5	20	13.21	15.08	129
3,5-diethyl-2-methyl-pyrazine	23.63	122, 135, **149**	6	20	13.09	12.99	177
Isopropenyl-pyrazine	29.02	67, 120, **119**	60	180	12.67	14.98	190
5H-5-Methyl-6,7-dihydrocyclopenta-pyrazine	30.18	119, 133, **134**	25	50	10.00	14.02	94

[1] Quantification ions are present in bold. [2] LOD and LOQ were obtained by standard solution of pyrazines, and precision was evaluated by commercial flavor rapeseed oil. [3] The concentration of pyrazines in the sample during the repeatability test. [4] Internal standards. Abbreviations: RSD—relative standard deviation.

This method allows the analysis of a variety of different pyrazines, and the selected internal standard is pyridine with a similar structure to pyrazines. Therefore, good selectivity is a necessary requirement for the correct identification and quantification of all analytes. The identification and quantitation of target compounds, including internal standard, were based on the retention time and different selected ions, which are listed in Table 1.

For the LOD and LOQ determinations, the refined rapeseed oil ("pyrazine-free") was used as the matrix to prepare standard solutions. The suitable concentration was established as LOD and LOQ of pyrazines with the signal-to-noise ratio (S/N) of pyrazines being 3.0 and 10.0, respectively [38]. As shown in Table 1, the LODs and LOQs of 13 pyrazines were in the range of 2–60 ng/g and 6–180 ng/g, respectively, indicating that MHS-SPME can be used in routine quantitation of pyrazines.

To evaluate the precision of the method, five parallel experiments were carried out on commercial flavor-enhanced rapeseed oils to calculate the intra-day precision. The inter-day precision was determined by analyzing the same sample five times a week for three weeks. Both intra-day precision and inter-day precision were expressed as the relative standard deviation (RSD). The results showed that the RSD of intra-day and inter-day were both lower than 16%, verifying the good precision of the MHS-SPME-arrow method. Model experiments were also carried out to evaluate accuracy.

Three spiked concentration levels in refined rapeseed oil were also studied by the standard addition method (SA). The results in Table 2 showed that the recoveries of pyrazines analyzed were in the range of 90% to 115% in all cases, indicating that the method was reliable and accurate within the concentration range of the model experiments.

Table 2. Recoveries of 13 pyrazines in refined rapeseed oil at different concentration levels.

Compound	Spiked Level (ng/g)	Recovery [1] (%)	Spiked Level (ng/g)	Recovery [1] (%)	Spiked Level (ng/g)	Recovery [1] (%)
2-methyl-pyrazine	342	104.4 ± 15.0	1710	106.1 ± 4.13	3420	98.5 ± 6.9
2,5-dimethyl-pyrazine	642	99.8 ± 7.7	3210	111.3 ± 13.0	6420	93.9 ± 12.5
2,6-dimethyl-pyrazine	382	109.8 ± 6.0	1910	104.3 ± 4.2	3820	100.9 ± 13.7
Ethyl-pyrazine	364	105.2 ± 2.9	1820	95.3 ± 9.5	3640	97.1 ± 14.4
2-ethyl-6-methyl-pyrazine	330	104.8 ± 8.4	1652	105.5 ± 6.5	3305	94.2 ± 10.7
2-ethyl-5-methyl-pyrazine	85	97.8 ± 8.3	425	93.4 ± 7.9	851	113.6 ± 5.5
Trimethyl-pyrazine	336	112.9 ± 3.1	1680	109.7 ± 5.6	3360	97.9 ± 10.6
2,6-diethyl-pyrazine	256	96.1 ± 6.5	1280	107.8 ± 8.1	2560	115.4 ± 7.2
3-ethyl-2,5-dimethyl-pyrazine	252	102.2 ± 6.7	1260	112.3 ± 15.7	2520	106.7 ± 7.9
2-ethyl-3,5-dimethyl-pyrazine	314	112.0 ± 14.0	1570	106.7 ± 6.3	3140	91.9 ± 9.4
3,5-diethyl-2-methyl-pyrazine	284	105.3 ± 10.3	1420	109.1 ± 6.23	2840	110.5 ± 6.5
Isopropenyl-pyrazine	308	98.7 ± 7.1	1540	95.34 ± 9.1	3080	93.0 ± 10.2
5H-5-Methyl-6,7-dihydrocyclopenta-pyrazine	98	96.5 ± 5.1	490	109.2 ± 6.7	980	111.7 ± 11.0

[1] Recovery calculated from three concentration and three replicates.

The stable isotope dilution analysis (SIDA) has proven to be very precise in the model experiments, even at a very low extraction yield [39]. The main reason is that SIDA uses the most suitable internal standard: Stable isotopes of the analytes, which can fully be recovered for the losses. This study compared the quantitative results of SIDA and MHS-SPME of flavor-enhanced rapeseed oil, using 2-methyl-pyrazine labeled $^2H_6$ as the stable isotope internal standard. The concentration of 2-methyl-pyrazine was 1.21 ± 0.13 μg/g for SIDA and 1.43 ± 0.11 μg/g for MHS-SPME-arrow, respectively. There was no difference (p >0.05) by statistical analysis. Thus, the data showed that the MHS-SPME-arrow could be applied to the analysis of these pyrazines in oil samples. By this conclusion, the method has been set up after condition optimization and SIDA method verification. An efficient and effective MHS-SPME-arrow-GCMS analytical protocol is established and can be used to quantify the flavor compounds in oils.

3.4. Analysis of Real Samples

Flavor-enhanced oils are usually produced by the traditional pressing process, mainly including roasting, pressing, and filtering. Heterocyclic volatiles are commonly produced by the Maillard reactions among proteins, amino acids, and sugars during oilseed roasting [40]. Pyrazines are one group of the main Maillard reaction products and also give the main source of roasting-like aroma in flavor oils [6]. Flavor-enhanced sesame and peanut oils are the main representatives of such oils, while rapeseed oil also has a good market in China for its unique spicy and roasted flavor, occupying more than 30% of the entire Chinese rapeseed oil market [41]. For the quantification of pyrazines in the three market product samples, MHS-SPME-arrow-GC-MS were carried out as set up above. Prior to analysis of the samples, the standard calibration solution of pyrazines and 3-methyl-pyridine with known concentrations were analyzed to calibrate the detector responses. Taking 2-ethyl-5-methyl-pyrazine in flavor-enhanced peanut oil as an example, Figure 3 showed the change of peak areas for four successive extractions. An exponential decrease in the target compound with the number of extractions was observed. The average concentrations and standard deviations of pyrazines together with R^2 and slope of the linear plot ln A_i versus i − 1, were shown in Table 3.

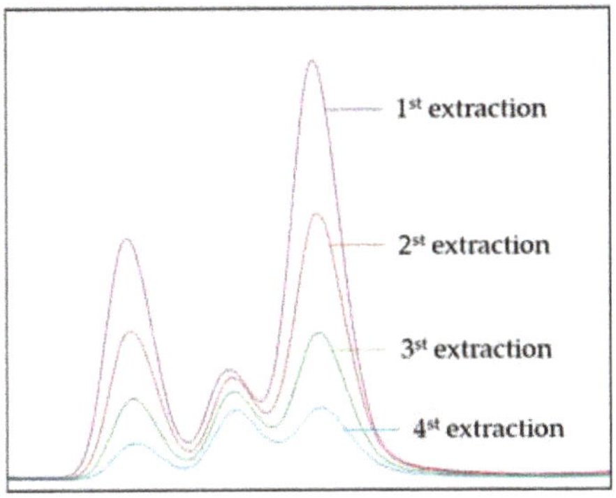

Figure 3. Chromatogram of 2-ethyl-5-methyl-pyrazine of four successive HS-SPME extractions from flavor-enhanced peanut oil.

Table 3. The coefficient of determination (R^2) and slope of the linear plot ln A_i versus i − 1, mean concentration, and standard deviation of pyrazines in peanut, sesame, and rapeseed oils.

	Peanut Oil			Sesame Oil			Rapeseed Oil		
	Concentration (mg/kg)	R^2	Slope	Concentration (mg/kg)	R^2	Slope	Concentration (mg/kg)	R^2	Slope
2-methyl-pyrazine	2.87 ± 0.18	0.9962	−0.71	15.14 ± 2.46	0.9943	−0.47	0.58 ± 0.00 [1]	0.9948	−0.56
2,5-dimethyl-pyrazine	3.23 ± 0.21	0.9990	−0.75	9.50 ± 1.01	0.9945	−0.56	1.07 ± 0.00	0.9979	−0.73
2,6-dimethyl-pyrazine	1.10 ± 0.07	0.9974	−0.74	3.98 ± 0.41	0.9937	−0.54	0.30 ± 0.02	0.9991	−0.65
Ethyl-pyrazine	0.38 ± 0.02	0.9973	−0.73	1.74 ± 0.12	0.9930	−0.49	0.09 ± 0.00	0.9932	−0.61
2-ethyl-6-methyl-pyrazine	0.76 ± 0.06	0.9998	−0.52	3.62 ± 0.71	0.9980	−0.41	0.24 ± 0.00	0.9991	−0.50
2-ethyl-5-methyl-pyrazine	0.96 ± 0.07	0.9999	−0.47	2.21 ± 0.46	0.9986	−0.41	0.21 ± 0.00	0.9988	−0.48
Trimethyl-pyrazine	1.07 ± 0.09	1.0000	−0.48	5.49 ± 1.16	0.9989	−0.38	0.29 ± 0.00	0.9995	−0.46
2,6-diethyl-pyrazine	0.06 ± 0.01	1.0000	−0.34	0.26 ± 0.06	0.9996	−0.27	0.02 ± 0.00	0.9997	−0.32
3-ethyl-2,5-dimethyl-pyrazine	0.83 ± 0.07	0.9998	−0.31	5.08 ± 1.22	0.9996	−0.25	0.55 ± 0.00	0.9999	−0.29
2-ethyl-3,5-dimethyl-pyrazine	0.12 ± 0.01	0.9997	−0.29	0.54 ± 0.13	0.9991	−0.24	0.03 ± 0.00	0.9897	−0.28
3,5-diethyl-2-methyl-pyrazine	0.05 ± 0.01	0.9965	−0.19	0.23 ± 0.05	0.9970	−0.16	0.02 ± 0.00	0.9959	−0.17
Isopropenyl-pyrazine	ND [2]	– [3]	–	0.93 ± 0.22	0.9994	−0.27	ND	–	–
5H-5-Methyl-6,7-dihydrocyclopenta-pyrazine	0.06 ± 0.01	0.9964	−0.21	0.47 ± 0.10	0.9862	−0.18	ND	–	–
Total pyrazines	11.51 ± 0.78			49.18 ± 3.25			3.42 ± 0.03		

[1] Standard deviation less than 0.005 was considered as 0.00. [2] ND, not detected. [3] Slope and R^2 not shown as it showed no linear decay.

Thirteen pyrazines were found in three oils. According to the discussion in Section 3.2, the optimal sample amount was taken as 50.0 mg. To apply the MHS-SPME method, the following requirements must be met: $R^2 > 0.75$ and $-0.9163 < \text{slope} < -0.0513$. The results showed that the R^2 of pyrazines in three samples was greater than 0.98, of which more than 90% was bigger than 0.99. All the slopes were in the range of −0.9163~−0.0513. It was shown that the ln A_i decreased linearly with the number of extractions, and the linear relationship was good, fitting to the MHS-SPME-arrow method.

There were significant differences in the concentrations of pyrazines in different types of oils ($p < 0.05$). The total concentrations of pyrazines in the three samples were, in order from high to low: Sesame oil, peanut oil, and rapeseed oil. The total amount of pyrazines in sesame oil was close to 14 times that of rapeseed oil. The pyrazines with the highest concentration were 2-methyl-pyrazine and 2,5-dimethyl-pyrazine in three oils. As described by previous studies [1,11], pyrazines may be the best indicator to measure roasted flavor intensity. It is now clear also that sesame oil has the strongest roasted flavor as the concentration of pyrazines is the highest.

4. Conclusions

In summary, a reliable MHS-SPME-arrow-GC-MS method combined with ISTD for the quantitation of 13 pyrazines in flavor oils was developed. The SPME-arrow fiber was found to be more suitable for the MHS-SPME method, and the PDMS/DVB/CAR fiber was selected. The highest efficiency was performed under selected HS-SPME conditions. The novel method was verified with the stable isotope dilution analysis method and showed high sensitivity and accuracy. Additionally, the pyrazines of three market product samples were analyzed and quantified using the new method, which proved that the MHS-SPME-arrow-GC-MS was suitable to quantify pyrazines in oils. In the future development, this method has great opportunities in the quantitation of other aroma active compounds such as alcohols and pyrroles. At the same time, as an absolute quantitative method to eliminate the matrix effect, this method also has an obvious limitation. That is that it requires multiple extractions for a single analysis, which means it takes a much longer time than the conventional HS-SPME-GC-MS. Therefore, there are some aspects for further improvement for this newly proposed method.

Author Contributions: Conceptualization, C.Z.; methodology, C.Z. and Z.X.; validation, C.Z. and Z.X.; investigation, Z.X. and C.Z.; project initiation and resources, X.X.; experimental work, Z.X.; writing—original draft preparation, Z.X.; writing—review and editing, C.Z., H.S., H.Z., Y.B., and X.X. All authors contributed to the project scope and content as well as solution discussion. All authors have read and agreed to the published version of the manuscript.

Funding: This research was funded by Wilmar (Shanghai) Biotechnology Research & Development Center Co., Ltd.; College of Food Science and Technology, Henan University of Technology.

Institutional Review Board Statement: Not applicable.

Informed Consent Statement: Not applicable.

Data Availability Statement: Data are contained within the article.

Conflicts of Interest: The authors declare no conflict of interest.

References

1. Liu, X.; Jin, Q.; Liu, Y.; Huang, J.; Wang, X.; Mao, W.; Wang, S. Changes in Volatile Compounds of Peanut Oil during the Roasting Process for Production of Aromatic Roasted Peanut Oil. *J. Food Sci.* **2011**, *76*, C404–C412. [CrossRef]
2. Guerra, P.V.; Yaylayan, V.A. Dimerization of Azomethine Ylides: An Alternate Route to Pyrazine Formation in the Maillard Reaction. *J. Agric. Food Chem.* **2010**, *58*, 12523–12529. [CrossRef]
3. An, A.; Kimpe, N.D. Formation of pyrazines from ascorbic acid and amino acids under dry-roasting conditions—ScienceDirect. *Food Chem.* **2009**, *115*, 1417–1423.
4. Kwon, T.Y.; Park, J.S.; Jung, M.Y. Headspace–Solid Phase Microextraction–Gas Chromatography–Tandem Mass Spectrometry (HS-SPME-GC-MS^2) Method for the Determination of Pyrazines in Perilla Seed Oils: Impact of Roasting on the Pyrazines in Perilla Seed Oils. *J. Agric. Food Chem.* **2013**, *61*, 8514–8523. [CrossRef] [PubMed]
5. Gama, A.P.; Adhikari, K. Sensory Characterization of Dominant Malawi Peanut Varieties After Roasting. *J. Food Sci.* **2019**, *84*, 1554–1562. [CrossRef]
6. Jia, X.; Zhou, Q.; Wang, J.; Liu, C.; Huang, F.; Huang, Y. Identification of key aroma-active compounds in sesame oil from microwaved seeds using E-nose and HS-SPME-GC×GC-TOF/MS. *J. Food Biochem.* **2019**, *43*, e12786. [CrossRef] [PubMed]
7. Qu, F.; Zhu, X.; Ai, Z.; Ai, Y.; Qiu, F.; Ni, D. Effect of different drying methods on the sensory quality and chemical components of black tea. *LWT Food Sci. Technol.* **2019**, *99*, 112–118. [CrossRef]
8. Baker, G.L.; Cornell, J.A.; Gorbet, D.W.; O'Keefe, S.F.; Talcott, S.T. Determination of Pyrazine and Flavor Variations in Peanut Genotypes During Roasting. *J. Food Sci.* **2010**, *68*, 394–400. [CrossRef]
9. Shimoda, M.; Shiratsuchi, H.; Nakada, Y.; Yin, W.; Osajima, Y. Identification and Sensory Characterization of Volatile Flavor Compounds in Sesame Seed Oil. *J. Agric. Food Chem.* **1996**, *44*, 3909–3912. [CrossRef]
10. Poehlmann, S.; Schieberle, P. Characterization of the Aroma Signature of Styrian Pumpkin Seed Oil (Cucurbita pepo subsp. pepo var. Styriaca) by Molecular Sensory Science. *J. Agric. Food Chem.* **2013**, *61*, 2933–2942. [CrossRef]

11. Zhou, Q.; Xiao, J.; Yao, Y.; Wang, B.; Huang, F. Characterization of the Aroma-Active Compounds in Commercial Fragrant Rapeseed Oils via Monolithic Material Sorptive Extraction. *J. Agric. Food Chem.* **2019**, *67*, 11454–11463. [CrossRef] [PubMed]
12. Kanavouras, A.; Kiritsakis, A.; Hernandez, R.J. Comparative study on volatile analysis of extra virgin olive oil by dynamic headspace and solid phase micro-extraction. *Food Chem.* **2005**, *90*, 69–79. [CrossRef]
13. David, F. 4.21—Application of Stir-Bar Sorptive Extraction in Food Analysis. In *Comprehensive Sampling and Sample Preparation*; Pawliszyn, J., Ed.; Academic Press: Oxford, UK, 2012; pp. 473–493. [CrossRef]
14. Cecchi, T.; Alfei, B. Volatile profiles of Italian monovarietal extra virgin olive oils via HS-SPME-GC-MS: Newly identified compounds, flavors molecular markers, and terpenic profile. *Food Chem.* **2013**, *141*, 2025–2035. [CrossRef]
15. Matheis, K.; Granvogl, M. Characterization of Key Odorants Causing a Fusty/Musty Off-Flavor in Native Cold-Pressed Rapeseed Oil by Means of the Sensomics Approach. *J. Agric. Food Chem.* **2016**, *64*, 8168–8178. [CrossRef] [PubMed]
16. Pawliszyn, J. *Handbook of Solid Phase Microextraction*; Elsevier: Waltham, MA, USA, 2012.
17. Pasqualone, A.; Paradiso, V.M.; Summo, C.; Caponio, F.; Gomes, T. Influence of Drying Conditions on Volatile Compounds of Pasta. *Food Bioprocess Technol.* **2014**, *7*, 719–731. [CrossRef]
18. Giannone, V.; Giarnetti, M.; Spina, A.; Todaro, A.; Pecorino, B.; Summo, C.; Caponio, F.; Paradiso, V.M.; Pasqualone, A. Physico-chemical properties and sensory profile of durum wheat Dittaino PDO (Protected Designation of Origin) bread and quality of re-milled semolina used for its production. *Food Chem.* **2018**, *241*, 242–249. [CrossRef] [PubMed]
19. Liu, X.; Wang, S.; Tamogami, S.; Chen, J.; Zhang, H. Volatile Profile and Flavor Characteristics of Ten Edible Oils. *Anal. Lett.* **2021**, *54*, 1423–1438. [CrossRef]
20. Song, N.-E.; Lee, J.-Y.; Lee, Y.-Y.; Park, J.-D.; Jang, H.W. Comparison of headspace–SPME and SPME-Arrow–GC–MS methods for the determination of volatile compounds in Korean salt–fermented fish sauce. *Appl. Biol. Chem.* **2019**, *62*, 1–8. [CrossRef]
21. Nam, T.G.; Lee, J.Y.; Kim, B.K.; Song, N.E.; Jang, H.W. Analyzing volatiles in brown rice vinegar by headspace solid-phase microextraction (SPME)–Arrow: Optimizing the extraction conditions and comparisons with conventional SPME. *Int. J. Food Prop.* **2019**, *22*, 1195–1204. [CrossRef]
22. Lee, J.-Y.; Kim, W.S.; Lee, Y.-Y.; Choi, Y.-S.; Choi, H.; Jang, H.W. Solid-phase microextraction Arrow for the volatile organic compounds in soy sauce. *J. Sep. Sci.* **2019**, *42*, 2942–2948. [CrossRef]
23. Manousi, N.; Rosenberg, E.; Zachariadis, G.A. Solid-Phase Microextraction Arrow for the Sampling of Volatile Organic Compounds in Milk Samples. *Separations* **2020**, *7*, 75. [CrossRef]
24. Kim, S.J.; Yoon, H.N.; Rhee, J.S. The effects of roasting temperatures on the formation of headspace volatile compounds in perilla seed oil. *J. Am. Oil Chem. Soc.* **2000**, *77*, 451–456. [CrossRef]
25. Zhang, W.; Rui, W.; Yuan, Y.; Yang, T.; Liu, S. Changes in volatiles of palm kernel oil before and after kernel roasting. *LWT Food Sci. Technol.* **2016**, *73*, 432–441. [CrossRef]
26. Ezquerro, Ó.; Pons, B.; Tena, M.A.T. Evaluation of multiple solid-phase microextraction as a technique to remove the matrix effect in packaging analysis for determination of volatile organic compounds. *J. Chromatogr. A* **2003**, *1020*, 189–197. [CrossRef]
27. Genovese, A.; Yang, N.; Linforth, R.; Sacchi, R.; Fisk, I. The role of phenolic compounds on olive oil aroma release. *Food Res. Int.* **2018**, *112*, 319–327. [CrossRef] [PubMed]
28. Pickard, S.; Becker, I.; Merz, K.H.; Richling, E. Determination of the alkylpyrazine composition of coffee using stable isotope dilution-gas chromatography-mass spectrometry (SIDA-GC-MS). *J. Agric. Food Chem.* **2013**, *61*, 6274–6281. [CrossRef]
29. Suzuki, M.; Tsuge, S.; Takeuchi, T. Gas chromatographic estimation of occluded solvents in adhesive tape by periodic introduction method. *Anal. Chem.* **1970**, *42*, 1705–1708. [CrossRef]
30. Mcauliffe, C.D. GC Determination of Solutes by Multiple Phase Equilibration. *Chem. Technol.* **1971**, *1*, 46–51.
31. Kolb, B. Multiple headspace extraction—A procedure for eliminating the influence of the sample matrix in quantitative headspace, gas chromatography. *Chromatographia* **1982**, *15*, 587–594. [CrossRef]
32. Ezquerro, Ó.; Pons, B.; Tena, M.A.T. Direct quantitation of volatile organic compounds in packaging materials by headspace solid-phase microextraction–gas chromatography–mass spectrometry. *J. Chromatogr. A* **2003**, *985*, 247–257. [CrossRef]
33. Gómez-Ariza, J.L.; García-Barrera, T.; Lorenzo, F.; Beltrán, R. Use of multiple headspace solid-phase microextraction and pervaporation for the determination of off-flavours in wine. *J. Chromatogr. A* **2006**, *1112*, 133–140. [CrossRef]
34. Canellas, E.; Vera, P.; Nerin, C. Multiple headspace-solid phase microextraction for the determination of migrants coming from a self-stick label in fresh sausage. *Food Chem.* **2016**, *197*, 24–29. [CrossRef] [PubMed]
35. Sgorbini, B.; Cagliero, C.; Acquadro, S.; Marengo, A.; Cordero, C.; Liberto, E.; Bicchi, C.; Rubiolo, P. Evaluation of volatile bioactive secondary metabolites transfer from medicinal and aromatic plants to herbal teas: Comparison of different methods for the determination of transfer rate and human intake. *J. Chromatogr. A* **2019**, *1594*, 173–180. [CrossRef] [PubMed]
36. Liu, G.; Xu, S.; Wang, X.; Jin, Q.; Xu, X.; Shen, Y.; Xu, G.; Zhang, H. Analysis of the volatile components of tea seed oil (Camellia sinensis O. Ktze) from China using HS-SPME-GC/MS. *Int. J. Food Sci. Technol.* **2016**, *51*, 2591–2602. [CrossRef]
37. Tena, M.T.; Carrillo, J.D. Multiple solid-phase microextraction: Theory and applications. *TrAC Trends Anal. Chem.* **2007**, *26*, 206–214. [CrossRef]
38. Martínez-Uruuela, A.; González-Sáiz, J.; Pizarro, C. Optimisation of the derivatisation reaction and subsequent headspace solid-phase microextraction method for the direct determination of chlorophenols in red wine. *J. Chromatogr. A* **2004**, *1048*, 141–151. [CrossRef]

39. Schieberle, P.; Grosch, W. Evaluation of the flavour of wheat and rye bread crusts by aroma extract dilution analysis. *Z. Lebensm. Unters. Forsch.* **1987**, *185*, 111–113. [CrossRef]
40. Van Lancker, F.; Adams, A.; De Kimpe, N. Formation of pyrazines in Maillard model systems of lysine-containing dipeptides. *J. Agric. Food Chem.* **2010**, *58*, 2470–2478. [CrossRef]
41. Zhang, Y.; Zhai, X.; Gao, L.; Jin, J.; Zhong, Q.; Sun, C.; Yan, L.; Liu, R.; Akoh, C.C.; Jin, Q. Quality of Wood-Pressed Rapeseed Oil. *J. Am. Oil Chem. Soc.* **2017**, *94*, 767–777. [CrossRef]

life

MDPI

Review

Preparation of Human Milk Fat Substitutes: A Review

Xuan Jiang [1,2], Xiaoqiang Zou [2,*], Zhonghao Chao [2] and Xiuli Xu [2]

1 Provincial Key Laboratory of Cereal Resource Transformation and Utilization, Henan University of Technology, Zhengzhou 450001, China; jiangxuan@stu.jiangnan.edu.cn
2 National Engineering Research Center for Functional Food, Collaborative Innovation Center of Food Safety and Quality Control in Jiangsu Province, National Engineering Laboratory for Cereal Fermentation Technology, State Key Laboratory of Food Science and Technology, School of Food Science and Technology, Jiangnan University, 1800 Lihu Road, Wuxi 214122, China; 6200113152@stu.jiangnan.edu.cn (Z.C.); 6200112092@stu.jiangnan.edu.cn (X.X.)
* Correspondence: xiaoqiangzou@jiangnan.edu.cn; Tel.: +86-510-8587-6799

Abstract: Human milk is generally regarded as the best choice for infant feeding. Human milk fat (HMF) is one of the most complex natural lipids, with a unique fatty acid composition and distribution and complex lipid composition. Lipid intake in infants not only affects their energy intake but also affects their metabolic mode and overall development. Infant formula is the best substitute for human milk when breastfeeding is not possible. As the main energy source in infant formula, human milk fat substitutes (HMFSs) should have a composition similar to that of HMF in order to meet the nutritional needs of infant growth and development. At present, HMFS preparation mainly focuses on the simulation of fatty acid composition, the application of structured lipids and the addition of milk fat globule membrane (MFGM) supplements. This paper first reviews the composition and structure of HMF, and then the preparation development of structured lipids and MFGM supplements are summarized. Additionally, the evaluation and regulation of HMFSs in infant formula are also presented.

Keywords: human milk; human milk fat substitutes; structured lipids; infant formula

Citation: Jiang, X.; Zou, X.; Chao, Z.; Xu, X. Preparation of Human Milk Fat Substitutes: A Review. *Life* **2022**, *12*, 187. https://doi.org/10.3390/life12020187

Academic Editors: Suzana Ferreira-Dias, Xuebing Xu and Fátima Peres

Received: 8 December 2021
Accepted: 24 January 2022
Published: 27 January 2022

1. Introduction

Human milk is the most ideal source of nutrition in the early stages of human life. It not only provides comprehensive and balanced nutrition for infants but also greatly reduces their incidence of diarrhea, allergic diseases and infectious diseases and reduces infant mortality [1,2]. Human milk fat (HMF) is one of the most complex natural lipids, with a unique fatty acid composition and distribution. HMF provides about 50% of the energy needed for the growth and development of infants [3]. HMF also plays an important role in the absorption of vitamins and mineral nutrients, as well as in the early neural development and physiological functions of infants [4]. In recent years, the rate of breastfeeding has been decreasing for a variety of reasons. Therefore, infant formula has become the best substitute for human milk. With the in-depth study of HMF and infant growth needs, the production of human milk fat substitutes (HMFSs) has aimed to simulate HMF composition, which can narrow the gap between breast-fed infants and formula-fed infants.

Since the development of the first HMFS, many innovations have been implemented to improve the nutrient intake and absorption of infant formula. The composition and structure of HMFSs affect the bioavailability of lipids in infants, hence affecting their growth and development. This article reviews the structural composition and nutritional characteristics of HMF and the research progress of HMFS preparation, and the evaluation and regulation of HMFSs are also summarized.

2. Physiochemical Properties of HMF

HMF is considered to be the reference or gold standard for the development of HMFSs [5–7]. Therefore, it is of great significance to study the composition of HMF.

Generally speaking, HMF accounts for 3–5% of human milk, mainly containing triacylglycerols (TAGs, 98–99%), phospholipids (0.26–0.80%), sterols (0.25–0.34%, mainly cholesterol) and trace amounts of various minor components, including monoacylglycerols (MAGs), diacylglycerols (DAGs), fatty acids and other substances [8]. These components are packaged into milk fat lipid globules, with phospholipids forming the bulk of the milk fat globule membrane (MFGM) and TAGs found in the core. The MFGM mainly consists of a complex mixture of (glyco)proteins (20–60%), triacylglycerols, glycerophospholipids (33% of the MFGM), sphingolipids and cholesterol [9]. Other than lipids, the MFGM is also composed of glycosylated peptides, filaments, mucin and lactadherin (Figure 1). These globules usually range from 1 to 10 µm across, with an average diameter of 4 µm in mature milk [10].

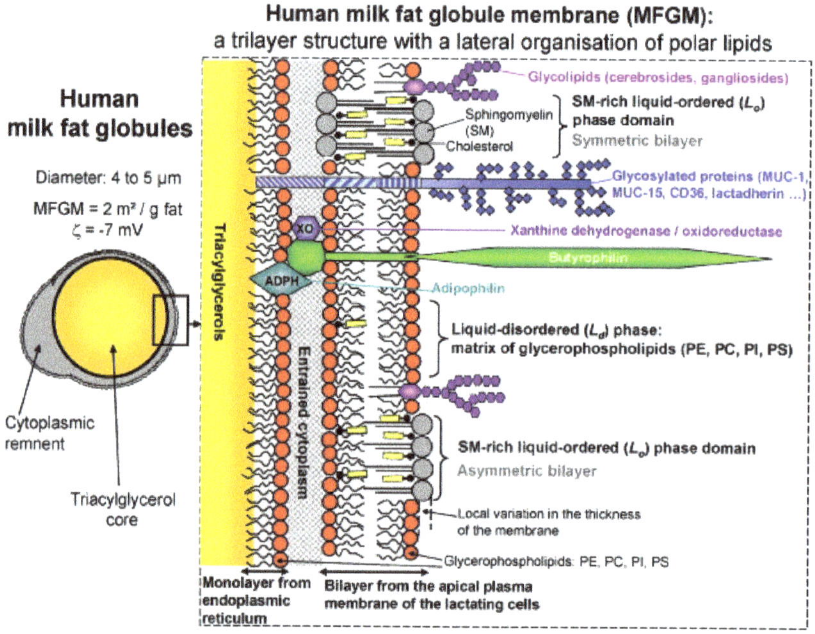

Figure 1. Structure of the milk fat globule and the membrane [11].

The human MFGM has a complex architecture in which polar lipids (PLs) are the backbone. According to the model described by Lopez [11], PLs are organized in the MFGM as a trilayer with the polar head groups exposed to an aqueous environment (cytoplasm, aqueous phase of milk) and the hydrocarbon tails forming a hydrophobic area in the center of the bilayer or in contact with the TAG core. At least two lipid phases coexist in the outer bilayer of the human MFGM: (1) the liquid-disordered phase composed of glycerophospholipids and (2) the lateral segregation of sphingomyelin (SM) in liquid-ordered phase domains. The kinked structure of the unsaturated fatty acyl chains of glycerophospholipids results in a shorter molecular length compared to straight SM molecules that have long-chain saturated fatty acids (SFAs) and a sphingoid base. Hence, there is probably an interdigitation among the tails of SM molecules in the MFGM bilayer, and cholesterol can fill the voids between the heads of SM.

2.1. *Fatty Acids and TAGs in HMF*

TAGs, the most abundant component of HMF, contain three fatty acids. The different structures (chain length, number and position of double bonds) and positions along the glycerol backbone of fatty acids lead to the complex composition of HMF. HMF contains more than 200 types of fatty acid [12]. However, only several fatty acids dominate, and other fatty acids are present in very low concentrations. Fatty acids in HMF have unique distribution characteristics and commonly appear in specific positions. Interestingly, the distribution of fatty acids in glycerol influences their availability [13]. Table 1 summarizes the total and sn-2 fatty acid composition of HMF. According to the saturation degree, fatty acids can be divided into SFAs, monounsaturated fatty acids (MUFAs) and polyunsaturated fatty acids (PUFAs).

Table 1. The total and sn-2 fatty acid composition of HMF [12–16].

Fatty Acid	Total Fatty Acid (%)	Sn-2 Fatty Acid (%)	Sn-2 Relative Percentage (%) *
C8:0	0.00–0.36	0.00–0.20	0.00–33.33
C10:0	0.15–3.10	0.21–1.60	8.25–33.33
C12:0	2.46–11.31	2.41–6.90	20.33–35.67
C14:0	2.46–11.63	6.20–15.40	37.58–57.97
C14:1	0.00–0.53	-	-
C15:0	0.09–1.11	0.46–0.53	72.09–78.47
C16:0	15.43–27.00	51.17–57.10	69.12–87.86
C16:1	0.00–3.60	1.60–3.50	14.81–38.99
C17:0	0.18–0.44	0.37–0.38	36.44–40.48
C17:1	0.10–0.34	-	-
C18:0	4.27–8.80	1.60–4.90	6.06–23.00
C18:1	28.30–45.88	8.10–17.43	7.89–14.18
C18:2	7.90–25.30	3.70–11.58	15.61–23.84
C18:3	0.00–1.50	0.28–0.90	15.15–26.30
C20:0	0.00–0.35	0.13–0.16	21.47–24.35
C20:1	0.23–1.68	0.40–0.51	17.41–22.06
C20:2	0.28–1.19	0.21–0.40	15.35–16.92
C20:3	0.25–1.57	0.25–0.34	16.66–20.28
C20:4	0.23–1.12	0.30–1.16	25.00–47.85
C20:5	0.00–0.24	-	-
C22:0	0.00–0.67	-	-
C22:1	0.00–0.66	-	-
C22:4	0.00–0.88	0.29–0.84	62.79–73.48
C22:5	0.00–0.22	0.27–0.33	66.92–72.29
C22:6	0.15–0.92	0.40–0.93	46.67–66.67

* Relative fatty acid in sn-2 position = (sn-2 fatty acid/3) × 100/total fatty acid in human milk. The sn-2 relative percentage data are calculated and concluded based on the data in the references.

The amount of fatty acids in HMF is not constant and depends mainly on the lactation period, feeding stage, diet and health of the infant [17–19]. In most HMF, the dominating fatty acids are SFAs and MUFAs [20–22]. The sum amount of SFAs and MUFAs is between 83% and 86%, of which approximately half are SFAs [23]. Among SFAs, palmitic acid (C16:0) content is the highest, followed by stearic acid (C18:0), myristic acid (C14:0) and lauric acid (C12:0) [24–26]. The pH of stomach acid in infants is higher than that in adults, and their resistance to foodborne pathogens is weaker. Lauric acid has high antibacterial activity, which is beneficial for the resistance of infants and young children to pathogenic bacteria [27]. In addition, odd-numbered carbon chain fatty acids (C15:0 and C17:0) are not common in natural fats and oils but are widely found in HMF. Among MUFAs, oleic acid (C18:1) is the most prevalent, constituting around 90% of MUFAs in human milk, and it also accounts for a high percentage in the total pool of fatty acids [23]. The distributions of fatty acids on TAGs are similar in most human milk, with most SFAs attached to the sn-2 position. HMF contains about 20% palmitic acid, with the majority attached to the sn-2 position [28,29]. The major unsaturated fatty acids, including oleic acid and linoleic acid, occupy the sn-1,3 positions [30].

HMF contains about 10–22% PUFAs, with linoleic acid (C18:2) accounting for the highest content. According to the position of the last double bond, PUFAs are divided into two types: n-3 and n-6 [31]. In the n-3 family, the precursor in this group, α-linolenic acid (C18:3n-3; ALA), also dominates. PUFAs with a chain length of more than 20 carbons are called long-chain polyunsaturated fatty acids (LC-PUFAs). It is worth noting that among the fatty acids at the sn-2 position, LC-PUFAs, such as docosatetraenoic acid (C22:4), eicosapentaenoic acid (C22:5, EPA) and docosahexaenoic acid (C22:6, DHA), account for the highest sn-2 relative percentage (Table 1).

Arachidonic acid (C20:4, ARA) is the most common n-6 LC-PUFA in HMF, while common n-3 LC-PUFAs include DHA, docosapentaenoic acid (C20:5, DPA) and EPA. Several kinds of LC-PUFAs, such as DHA and ARA, exist at high concentrations in human central nervous tissue, especially in some specific brain regions and retinal cell membranes [32,33]. Their rapid accumulation in infants mainly occurs in the last three months of pregnancy and the first two years after birth, which is the period of rapid brain development in infants [34,35]. LC-PUFAs are important components of the neural cell membrane. It has been proved that LC-PUFAs can promote brain development by promoting neural stem cell differentiation, promoting synaptic formation and improving synaptic transmission mechanisms [36]. The postpartum period is the key period for the development of immune function in newborns. Research has shown that LC-PUFAs in HMF also play an important role in the improvement of the infant immune system [37,38]. As is shown in Figure 2, DHA and ARA can be synthesized in vivo through the action of elongase, desaturase and peroxisome on linoleic acid and ALA [39]. However, the enzyme system in infants is not well developed. Infants cannot synthesize enough LC-PUFAs to meet the requirements for growth, and thus, dietary supplements are prerequisite and vital.

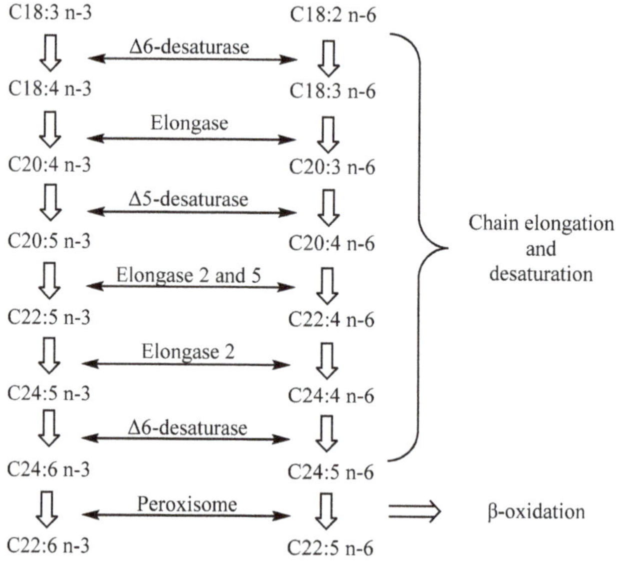

Figure 2. The flow chart of the synthesis of LC-PUFAs from linoleic acid and ALA through desaturation, carbon chain elongation and β-oxidation in vivo.

2.2. *Polar Lipids in HMF*

In addition to neutral lipids, HMF also contains PLs, which are mainly distributed in the MFGM. PLs mainly include phosphatidylinositol (PI), phosphatidylcholine (PC), phosphatidylethanolamine (PE), phosphatidylserine (PS) and SM. Other PLs, such as lysophospholipids and plasmalogens, have also been found in human milk [40]. Despite their low content, PLs are important structural components due to their nutritional properties. Generally speaking, SM (range 7.96–16.49 mg/100 mL) is the most abundant PL in HMF, followed by PE (6.49–15.90 mg/100 mL), PC (3.63–8.09 mg/100 mL), PS (3.25–6.74 mg/100 mL) and PI (3.07–6.32 mg/100 mL) [11].

Table 2 summarizes the fatty acid composition of different types of PLs in HMF. Generally speaking, PC, PI and PS exhibit similar distributions of SFAs and unsaturated fatty acids (UFAs, around 50%). The UFA content in PE accounts for around 65%. However, SM presents a high SFA distribution of 70–80%, which helps to reduce liquidity and maintains the rigidity and structure of the MFGM [41]. Palmitic acid is the main fatty acid in PC, while stearic acid is the most abundant fatty acid in PE, PI and PS. Additionally, behenic acid (C22:0) is the main fatty acid in SM. ARA, EPA and DHA, which are vital to the growth and brain development of newborns, are mainly distributed in PE, PI and PS.

Table 2. Fatty acid composition of PLs in HMF (%) [42].

Fatty Acid	PE	PI	PS	PC	SM
C10:0	-	-	-	-	0.1
C12:0	0.1–0.6	0.2–1.2	0.1–1.2	0.1–0.4	0.2–0.6
C14:0	0.2–2.4	0.4–3.3	0.1–2.5	0.9–4.5	1.1–2.1
C15:0	0.1–0.2	0.1–0.7	0.1–0.2	0.2–0.4	0.1–0.8
C16:0	7.2–11.8	5.8–17.3	7.3–13.4	25.1–38.0	5.3–21.3
C16:1	0.5–2.4	0.2–2.1	0.6–2.0	0.4–1.7	0.1–0.7
C17:0	0.2–1.5	0.2–0.7	0.6–1.0	0.3–0.7	0.5–1.4
C17:1	-	-	-	-	0.3
C18:0	23.1–29.1	30.6–34.5	33.5–42.8	16.9–24.7	11.8–13.8
C18:1	15.8–23.7	12.4–20.1	15.7–19.4	14.0–20.8	1.0–4.0
C18:2	13.0–23.8	5.3–19.5	8.5–23.0	13.9–24.1	0.3–4.5
C18:3	0.2–4.1	0.1–2.5	0.1–2.4	0.2–1.3	0.1–0.7
C19:0	-	-	-	-	0.4
C20:0	0.3–0.4	0.5	0.5	0.25–0.3	6.4–10.9
C20:1	1.3–1.4	0.2	0.5	0.4–0.7	0.1–0.5
C20:2	0.3–1.1	0.2–0.8	0.2–1.4	0.1–0.3	0.6
C20:3	1.1–3.5	2.0–5.2	1.3–3.9	0.6–2.4	0.2–03
C20:4	4.8–12.7	4.5–12.2	1.5–4.6	1.7–3.3	0.3–0.5
C20:5	0.3–4.2	11.7	0.5–9.0	0.1–2.9	0.2–5.3
C21:0	-	-	-	-	0.8–2.6
C22:0	0.2	-	-	0.2	12.9–20.7
C22:1	0.1–0.2	0.4	0.5	0.1–0.3	0.4–11.8
C22:2	1.5	-	-	0.1	4.8
C22:4	2.1–3.9	1.4–6.0	1.4–4.2	0.3–0.7	-
C22:5	0.8–2.4	0.4–2.2	1.6–3.0	0.4–0.9	-
C22:5	0.7–2.3	0.1–0.7	0.5–0.9	0.1–0.2	0.1
C22:6	1.0–5.1	0.4–1.7	1.5–2.9	0.1–0.6	0.5–1.1
C23:0	-	-	-	-	4.0–7.7
C24:0	0.3–2.8	0.9	1.2	0.1–0.5	8.1–19.5
C24:1	0.1–0.2	0.5	0.5	0.1–0.7	9.7–17.7

PLs derived from HMF have biological characteristics. As the precursors of intracellular messengers, PC and SM contents are considered to be very important for the growth and development of infants. For newborns and young children, PC and SM are the main sources of choline. About 17% of the total exogenous choline in newborns comes from these PLs [43]. Choline is an important nutrient that participates in a variety of biological processes, such as metabolism. It also exists in the membrane structure of the brain and nerve tissue [44]. Choline is the precursor of the neurotransmitter acetylcholine, which can regulate transduction signals and be a source of methyl groups in intermediate metabolism [45]. Choline is secreted from the maternal circulation into human milk and increases during lactation. The intact SM and its metabolites are important for the development of colon tumors, as well as the inflammatory process, cell function and growth [46–48].

Sphingolipids in the MFGM are believed to prevent globules from coalescing and to be critical for the formation and maintenance of the globular membrane structure [49,50]. The sphingolipids found in human milk include sphingomyelin and glycosphingolipids (glycolipids), and sphingomyelin (SM) is the most predominant species. Glycosphingolipids, which are quantitatively minor constituents of the MFGM, comprise cerebrosides (neutral glycosphingolipids containing uncharged sugars) and gangliosides (acidic glycosphingolipids containing sialic acid) [51]. The major ganglioside in HMF is monosialoganglioisde (GM_3). Human milk GM_3 is rich in saturated fatty acids, including palmitic acid (C16:0), stearic acid (C18:0), arachidic acid (C20:0), behenic acid (C22:0) and lignoceric acid (C24:0) [52]. Gangliosides provide sialic acid and have roles in immune protection in the infant by acting as prebiotic factors, as well as decoy receptors, or competing with pathogens for receptor sites on intestinal epithelial cells [53,54].

2.3. Digestion and Absorption of HMF

As shown in Figure 3, TAGs in HMF undergo preliminary hydrolysis under the catalysis of gastric lipase. Gastric lipase is an sn-1,3-specific enzyme that preferentially hydrolyzes fatty acids at the sn-3 position and then generates free fatty acids (FFAs) and DAGs. Gastric lipase in infants can only hydrolyze 10–30% of TAGs [55]. DAGs have an emulsifying effect and help to further digest TAGs in the small intestine. Then, the initially digested HMF enters the small intestine and is hydrolyzed into sn-2 MAGs and FFAs under the catalysis of pancreatic lipase. Sn-2 MAGs are further transported to the brush border of the small intestine mucosa and absorbed by small intestinal epithelial cells [56].

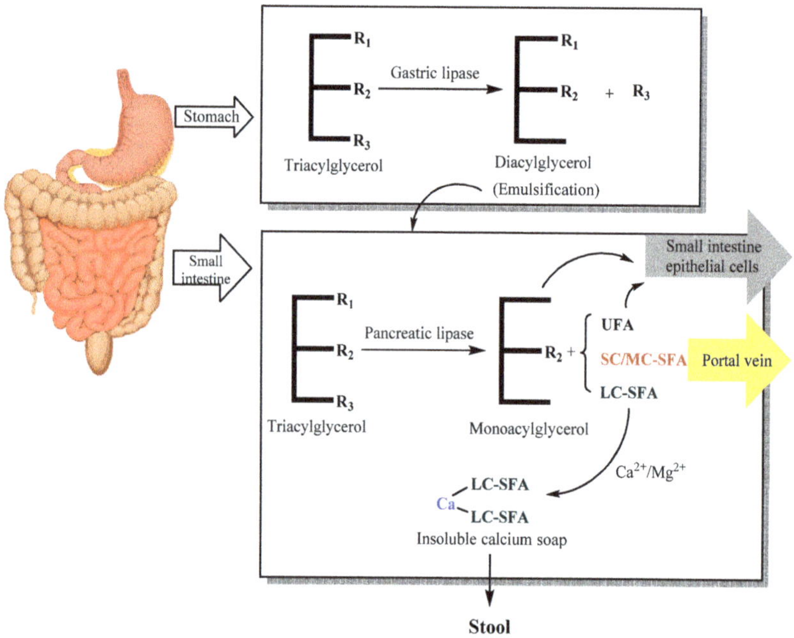

Figure 3. The digestion process of TAGs in infants. (R_1–R_3: random fatty acid; UFA: unsaturated fatty acid; SC/MC-SFA: short-/medium-chain saturated fatty acid; LC-SFA: long-chain saturated fatty acid).

The absorption of TAG hydrolysates is related to the composition and positional distribution of fatty acids [57]. The absorption of FFAs is affected by the chain length and degree of unsaturation [58]. SFAs with shorter carbon chains are transported to the liver for oxidation through the portal vein, while SFAs with longer carbon chains cannot pass through the portal vein [59,60]. In the environment of the small intestine in infants, free long-chain SFAs have a higher melting point than UFAs. They can easily combine with calcium and other metal ions to form insoluble soaps, causing the loss of fatty acids and calcium, as well as hard stools or even constipation [61]. Therefore, long-chain saturated fatty acids are absorbed more easily in the form of sn-2 MAGs [62].

As one of the most important fatty acids in HMF, palmitic acid is specifically distributed on TAGs, with more than 70% located at the sn-2 position [28,29]. Palmitic acid at the sn-2 position is more conducive to being absorbed and utilized [63,64]. Lipid digestion mainly relies on pancreatic lipase, which is sn-1,3-specific. Palmitic acid at the sn-1,3 position turns into FFAs after hydrolysis and binds with intestinal calcium and magnesium ions, generating insoluble fatty acid soap salt [65]. This can lead to low efficiency of fatty acid intake, loss of minerals and even stool sclerosis, constipation and other digestive diseases. In contrast, palmitic acid at the sn-2 position was proved to be absorbed much more efficiently [14]. Furthermore, palmitic acid at the sn-2 position plays an important role in improving the growth rate and the mineral density of bone, reducing infant crying and promoting the growth of intestinal probiotics [66,67].

At the same time, medium-chain fatty acids (MCFAs) are also important components in HMF, accounting for about 8–10%. MCFAs are easier to absorb than long-chain fatty acids (LCFAs). At present, manufacturers normally use medium-chain triacylglycerols (MCTs) to supplement MCFAs in infant formula. However, high levels of dietary MCTs may result in an undesirable increase in the level of circulating dicarboxylic acids, which can cause excessive ketone body poisoning [68]. It was found that the majority of MCFA-containing TAGs in HMF are medium- and long-chain triacylglycerols (MLCTs) rather than MCTs [69]. In addition, MCFAs can also limit the oxidation of PUFAs and LC-PUFAs and enhance the conversion of PUFAs to LC-PUFAs [70]. In MLCTs, MCFAs are released into the blood at a more stable rate of hydrolysis, avoiding excessive ketone body poisoning, and can also improve the absorption of LCFAs. MLCTs can also effectively reduce blood lipid and cholesterol levels, preventing and treating blood clots and other vascular diseases [71,72].

3. Production of HMFSs

3.1. Blending

Blending refers to the process of simply blending different animal fats, vegetable oils or microbial oils to prepare HMFSs. It has the advantages of simple operation and lack of by-products, and it can be used in large-scale commercial production. The development of HMFSs began in the 19th century, with the goal of mimicking the lipid content of HMF. Most infant foods of the first generation, represented by the first commercial infant formula developed by Leibig, had high total calories from protein and low values from fats, among which bovine milk fat (BMF) was applied as HMFSs [73,74].

At the beginning of the 20th century, the difference in fatty acid composition between BMF and HMF was clearly recognized. Hence, the composition of infant formula underwent "humanization" transformation. Gerstenberger developed blends of tallow, tallow oil, coconut and cocoa butter to mimic HMF. Based on the chemical analysis available at the time, it was determined that this mixture of fats was nearly identical to HMF in terms of average molecular weight, saturation and proportion of fatty acid [75]. Cod liver oil was used in some formulas decades before the important role of DHA in infant development was discovered [76].

At present, most commercial infant formulas still contain HMFSs prepared by pure blending. The main types are as follows: 1. A mixture of single vegetable oil and milk fat; 2. A mixture of various vegetable oils and milk fats; 3. A mixture of vegetable oils. Generally, the combination of commercial vegetable oils (coconut oil, palm oil, high oleic safflower oil or sunflower oil, soybean and rapeseed oil) is a good source of SFAs (C8:0–C18:0). Simple mixtures of unsaturated vegetable oils (e.g., corn or soy oil) and fats that contain a predominance of lauric acid or shorter saturated fatty acids (e.g., coconut oil) are common fat blends in some formulas. Complex mixtures of oils that contain modest levels of long-chain SFAs can be used in developing formulas with fatty acid profiles closer to that of human milk [57]. MCTs, predominantly containing caprylic acid (C8:0) and capric acid

(C10:0), have been added to some preterm formulas to maximize fatty acid absorption in preterm infants [77].

Approximately 70% of palmitic acid in human milk is in the sn-2 position, whereas this fatty acid is primarily in the sn-1 and sn-3 positions in most animal and vegetable fats. Lard contains approximately 80% of its long-chain SFAs in the sn-2 position, which is similar to the SFA content in HMF. Therefore, several studies have used lard-based formulas and proved their efficient fat absorption [63,64]. However, there are numerous cultural and religious prohibitions against using lard. In the late 20th century, multiple studies used blends of palm olein and other vegetable oils in different ratios to achieve proportions of palmitic and oleic acid that were similar to those in HMF [78–81]. Through comparison, they proved that infant formulas with palm olein at levels similar to HMF lower the absorption of calcium and/or fat with increased stool hardness and lower bone mass.

PUFAs (especially DHA and ARA) are supplemented from docosahexaenoic acid single cell oil (DHASCO), arachidonic acid single cell oil (ARASCO) and marine oil [82,83]. The fatty acid compositions of commonly used oils are shown in Tables 3 and 4. Fish oil generally contains DHA, EPA and small amounts of ARA and DPA, and their composition varies depending on diet, location, season and physiological conditions (such as species age and gender) [84]. Recently, it was found that the reason for the high content of LC-PUFAs in marine fish oil is that marine fish feed on plankton consuming microalgae, which have the ability to synthesize and accumulate a large number of LC-PUFAs [85]. In addition, some microorganisms, such as bacteria and fungi, have also been proved to have the ability to synthesize LC-PUFAs. A study confirmed that LC-PUFAs derived from single cells have no adverse effects on the growth and overall health of infants [86]. At the same time, it is necessary to ensure reasonable proportions of PUFAs during blending. An imbalance between n-3 and n-6 LC-PUFAs will affect the development of language function and the growth and development of children [87]. Additionally, different LC-PUFAs should be reasonably added during the preparation of HMFSs. An imbalance of dietary DHA and ARA may have negative impacts on cognitive development [88,89].

Table 3. Fatty acid composition of bovine milk fat and vegetable oils (%) [90–93].

Fatty Acid	Bovine Milk Fat	Coconut Oil	Palm Oil	Safflower Oil	Sunflower Oil	Soybean Oil	Canola Oil
C4:0	4.00–5.10	-	-	-	-	-	-
C6:0	2.10–2.90	0.00–0.04	-	-	-	-	-
C8:0	1.20–1.90	5.80–7.00	-	-	-	-	-
C10:0	2.40–3.50	4.80–8.00	-	-	-	-	-
C10:1	0.20–0.40	-	-	-	-	-	-
C12:0	3.00–4.10	48.00–51.02	-	-	-	-	-
C14:0	10.00–12.10	16.00–21.80	1.23–1.70	0.00–0.50	-	0.00–0.50	0.00–0.06
C14:1	0.40–1.30	-	-	-	-	-	-
C15:0	0.80–1.10	-	-	-	-	-	-
C16:0	28.70–34.10	8.40–9.20	41.78–43.30	4.00–7.50	3.70–6.90	9.00–14.50	3.75–10.50
C16:1	0.12–2.20	-	-	-	-	-	0.00–0.21
C17:0	0.40–0.50	-	-	-	-	-	0.00–0.04
C17:1	0.10–0.30	-	-	-	-	-	-
C18:0	10.30–13.30	1.94–2.80	3.39–4.80	2.50–2.70	1.98–2.90	4.00–5.20	1.87–6.90
C18:1	21.70–28.00	5.84–8.80	41.90–42.40	16.60–18.70	31.50–45.39	25.40–45.39	23.20–62.41
C18:2	1.50–2.30	0.50–1.28	7.80–11.03	71.10–76.00	46.02–59.50	46.02–51.90	15.20–20.12
C18:3	0.90–1.40	-	-	-	0.00–0.12	0.12–8.00	8.37–44.00
C20:0	0.20–0.20	0.00-0.25	-	0.00–0.20	0.00–2.30	-	0.00–0.64
C20:1	-	-	-	-	-	-	0.00–1.54
C20:2	-	-	-	-	-	-	0.00–0.11
C22:0	-	-	-	-	-	-	0.00–0.35
C24:0	-	-	-	-	-	-	0.00–0.27
C24:1	-	-	-	-	-	-	0.00–0.26

Table 4. Fatty acid composition of sources of LC-PUFAs used in infant formula (%) [94–99].

Fatty Acid	ARASCO	DHASCO	Tuna Oil	Cod Liver Oil
C8:0	-	-	-	-
C10:0	-	-	-	-
C12:0	-	3.60–4.40	-	2.21
C13:0	-	-	-	-
C14:0	0.34–0.58	18.50–19.40	3.27–3.42	3.83
C14:1	-	-	0.00–0.14	-
C14:2	-	-	-	-
C15:0	-	-	0.00–1.06	-
C15:1	-	-	0.00–0.09	-
C16:0	7.17–9.59	18.00–18.10	15.78–20.73	10.60
C16:1	-	1.80–2.00	4.14–6.14	6.97
C16:2	-	-	-	1.02
C17:0	-	-	1.39–1.58	-
C17:1	-	-	0.00–0.79	-
C18:0	7.70–10.50	0.40–1.00	4.52–5.89	2.73
C18:1	14.00–23.35	15.00–15.40	16.32–19.35	19.40
C18:2	4.56–7.62	0.60–1.00	1.35–1.84	1.43
C18:3	2.45–4.00	-	0.76–3.94	1.27
C18:4	-	-	0.00–1.23	2.29
C20:0	0.00–0.96	-	-	-
C20:1	-	-	0.00–1.76	9.40
C20:2	-	-	0.00–0.25	0.53
C20:3	0.00–4.30	-	0.00–0.31	0.47
C20:4	42.69–48.50	-	2.49–3.89	1.03
C20:5	-	-	6.35–7.62	8.89
C22:0	0.00–2.02	-	-	-
C22:1	-	-	0.00–0.93	7.57
C22:2	-	-	0.00–0.13	-
C22:4	-	-	0.00–1.20	0.50
C22:5	-	-	1.57–2.84	1.13
C22:6	-	38.40–39.00	22.85–26.86	10.70
C24:0	1.30–2.04	-	-	1.32
C24:1	-	-	0.00–0.77	-

Some studies have proved that HMFSs in commercial milk powder can imitate HMF at the fatty acid level [100]. However, the blending method can only change the fatty acid composition and content of HMFS. The distribution of fatty acids in traditional HMFSs is far from that of HMF, and thus, it cannot meet all of the nutritional needs of infants. Therefore, the preparation of HMFSs is not limited to unmodified oil but has started to include structured lipids (SLs) as raw materials [101,102].

3.2. Preparation of SLs

In 1968, Wyeth published the results of an animal study showing that a blend with a high proportion of sn-2 palmitate oil was well absorbed and was better than other blends, including butterfat, lard and vegetable oils [103]. These results offered an explanation for the superior absorption of HMF that could not be explained by fatty acid composition alone. The importance of the positional distribution of fatty acids has been highly valued. Hence, the preparation of SLs has attracted much attention. The main methods of SL preparation include the chemical method and the enzymatic method. Although chemical synthesis is an important process in the industry, it has many disadvantages in general, such as high reaction temperature, weak specificity, many reaction by-products, poor product flavor, high requirements of equipment, etc. Therefore, many researchers have focused on enzymatic synthesis. The enzymatic synthesis method has many advantages, such as mild reaction conditions, simple separation and purification steps, and strong product specificity, which can synthesize HMFSs with special structures to meet the needs of infant growth and nutrition [104].

3.2.1. Enzymatic Acidolysis

Acidolysis is the most commonly used method for the preparation of SLs [105]. As is shown in Figure 4, enzymatic acidolysis is a method in which TAGs exchange acyl groups with fatty acids under the catalysis of lipase. During the acidolysis process, fatty acids at the sn-2 position easily migrate to the sn-1,3 positions in the intermediate by-product (DAGs), which ultimately leads to a decrease in the purity of the target product [106]. This method has the advantages of simple operation, fewer kinds of products and easy separation, which makes it the most commonly used synthetic method at present [107].

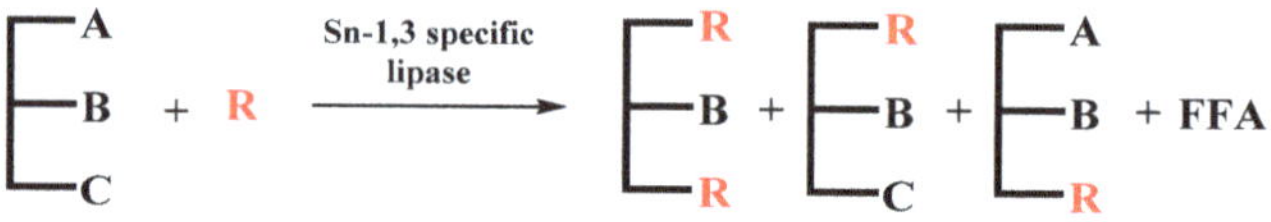

Figure 4. Acidolysis process route of structured lipids (A–C: original fatty acids; R: random fatty acid).

Preparation of sn-2 Palmitate SLs through Acidolysis

The acidolysis reaction is mainly applied in the preparation of sn-2 palmitate SLs. Most studies have used palmitic acid–rich sources, among which the most common one is tripalmitin [107]. Ilyasoglu et al. [108] synthesized HMFSs by Lipozyme® RM IM (*Rhizomucor miehei* lipase)-catalyzed acidolysis among tripalmitin, hazelnut oil fatty acids and Neobee® fatty acids. The SLs contained caprylic acid (12.8 g/100 g), capric acid (10.6 g/100 g) and palmitic acid (30 g/100 g) [106].

On the other hand, other researchers have used animal fat as the raw material to produce HMFSs. Lard is a rich renewable resource in China and contains 29.5% palmitic acid and 74% at the sn-2 position, which is similar to human milk [109]. However, this renewable resource cannot be effectively used because lard has a high melting point and mainly contains LCTs, which affects its physical, chemical and physiological properties and limits its application in the food industry. In the study of Yang et al. [110], lard and fatty acids from soybean were esterified in a solvent-free system at both the laboratory scale and large scale, producing HMFSs containing 71.1% sn-2 palmitic acid and 44.4% sn-1,3 oleic acid. Zhao et al. [111] synthesized SLs that incorporated 34.2 mol% capric acid through acidolysis between lard and capric acid in a hexane system. Nielsen et al. [112] produced HMFS through acidolysis of lard and fatty acid from soybean oil. The content of sn-2 palmitic acid in the product accounted for 71.9%, and the content of linoleic acid and linolenic acid increased from 9.2% and 0.8% to 23.8% and 2.3%, respectively. Basa catfish oil is a newly discovered natural oil whose fatty acid composition and distribution are similar to those of HMF. Zou et al. [113] prepared SLs containing 58.43% sn-2 palmitic acid and 80.16% sn-1,3 oleic acid through a two-step process from basa catfish oil. The process includes fractionation to enrich TAG fractions with a high content of sn-2 palmitic acid and Lipozyme® RM IM-catalyzed acidolysis of the fractionated products with FFAs from high-oleic-acid sunflower oil to increase the content of 1,3-oleoyl-2-palmitoylglycerol (OPO). However, consumers may be reluctant to use animal fat to feed infants. Therefore, in recent years, there have been studies using algal oil rich in palmitic acid [114].

In addition to natural lipids, some studies have used a two-step acidolysis method to prepare SLs in order to increase the content of sn-2 palmitic acid. Jiménez et al. [115,116] first obtained FFAs rich in palmitic acid through saponification and fractionation and then used an acidolysis reaction between FFAs and palm stearin to prepare SLs with an sn-2 palmitic acid content of up to 78.5%. These palmitic acid–rich TAGs can be used to obtain structured TAGs rich in palmitic acid at the sn-2 position and oleic acid at sn-1,3 positions. Esteban et al. [117] prepared palmitic acid–enriched acylglycerols with palmitic acid and palm stearin through two-step acidolysis and then used oleic acid–enriched FFAs for further enzymatic acidolysis. In the solvent and solvent-free systems, SLs with up to 67.2% oleic acid at sn-1,3 positions and 67.8% palmitic acid at the sn-2 position were obtained under the applied conditions.

Preparation of SLs Rich in LC-PUFAs through Acidolysis

LC-PUFAs are important components of HMFSs due to their vital influence on growth and development. SLs rich in LC-PUFAs are often prepared through the acidolysis process. Sahin et al. [118] prepared fatty acid and fatty acid ethyl ester by hydrolysis and alcoholysis of DHASCO and ARASCO, respectively. The FAEEs were used as acyl donors for acidolysis and transesterification with tripalmitin. After purification, the content of sn-2 palmitic acid in the product reached 48.53%, and the contents of DHA and ARA were 10.75% and 17.69%, respectively. Through a similar process, Li et al. [119] enriched refined olive oil with palmitic acid and DHA through acidolysis. In the product, palmitic acid was incorporated into the TAGs of refined olive oil at 55.79 mol%. Meanwhile, the incorporated palmitic acid at the sn-2 position and total DHA were found to be up to 33.63 mol% and 3.54 mol%, respectively. Pande et al. [120] used tripalmitin, extra virgin olive oil FFAs and DHASCO FFAs as the substrates of acidolysis. In their product, sn-2 palmitic acid accounted for 60 mol%, and the oleic acid content in sn-1,3 positions was predominant, while the DHA content accounted for 5.89–7.54% in different SLs. In the study of Nagachinta and Akoh [121], DHA and γ-linolenic were incorporated into SL products to improve their nutritional value while maximizing the content of palmitic acid at the sn-2 position. The SL product contained 35.11% palmitic acid at the sn-2 position, with 3.75% DHA and 5.03% γ-linoleic acid. Nagachinta and Akoh [122] synthesized SLs containing 56.9 mol% sn-1,3 ARA and 28.7 mol% sn-2 palmitic acid through enzymatic acidolysis between ARA and tripalmitin.

Though SLs rich in sn-1,3 LC-PUFAs have a positive impact on lipid metabolism [123], considering the physiological function and stability of the sn-2 position LC-PUFAs, recent studies have explored the preparation of SLs rich in LC-PUFAs at the sn-2 position. For example, Zou et al. [124] and Naranjo et al. [125] prepared SLs containing 87.45 and 54.4 mol% sn-2 LC-PUFAs through acidolysis between microbial oil and MCFAs, respectively.

3.2.2. Enzymatic Interesterification

As is shown in Figure 5, interesterification mainly refers to the enzymatic reaction between different TAGs or between TAGs and fatty acid ethyl ester. Due to its low specificity, low synthetic efficiency and difficult separation of reaction products, interesterification, especially reactions between TAGs, is not widely used. Enzymatic interesterification is a directional process catalyzed by lipase. Depending on enzyme regioselectivity, enzymatic interesterification can be classified as random or specific. Intermediate specificity can also be obtained simply by adjusting the residence time of the enzymatic process [125]. Due to the high selectivity of lipase, enzymatic interesterification only produces a few unwanted by-products, leading to fewer post-treatment products. Enzymatic interesterification is carried out under mild operating conditions, resulting in a lower level of PUFA degradation.

Maduko and Park [126] modified the structure of TAGs in a vegetable oil mixture under the catalysis of Lipozyme® RM IM through interesterification between the vegetable oil mixture and tripalmitin. These authors studied the incorporation of palmitic acid, oleic acid and linoleic acid at the sn-2 position and obtained a product with a similar fatty acid profile to that of HMF. In the product, MLCTs with palmitic acid at the sn-2 position accounted for 64–66%. Korma et al. [127] used the interesterification of ARASCO and MCTs in a solvent-free system to prepare SLs with an MLCT content of 53.75%. In the research of Ghosh et al. [128], the content of palmitic acid in TAGs was increased by fractionated palm stearin. The product of interesterification between palm stearin fractionate and fish oil contained 75.98% palmitic acid at the sn-2 position, as well as 0.27% ARA, 3.43% EPA and 4.25% DHA. Bektas et al. [129] used ethyl oleate as the acyl donor, which greatly increased the acyl group incorporation rate, and the sn-1,3 oleic acid content reached 64.9%, while its sn-2 palmitic acid content remained at around 80.6%. Rodriguez and Akoh [130] synthesized sn-2 palmitic acid–rich SLs through the interesterification reaction of amaranth oil, DHASCO and ethyl palmitate and then prepared HMFSs containing 33.9% palmitic acid and 1.9% DHA.

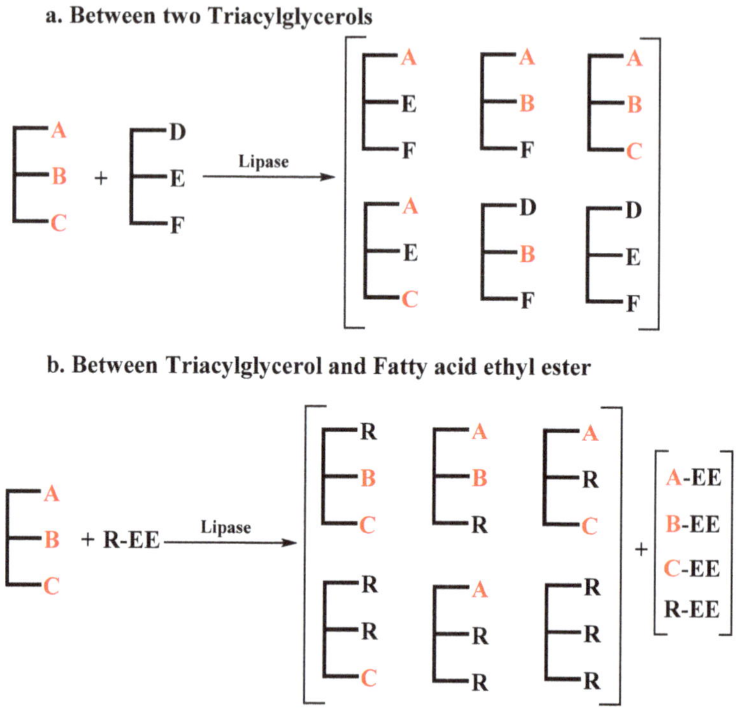

Figure 5. Process route of enzymatic interesterifications between two triacylglycerols (**a**) and between triacylglycerol and fatty ethyel ester (**b**) (A–F: fatty acid; R: random fatty acid; A-C/R-EE: fatty acid ethyl ester).

3.2.3. Enzymatic Esterification

There are only a few studies on esterification in SL production. As is shown in Figure 6, esterification is a reaction between fatty acids and glycerol under the catalysis of lipase. It has the advantage of enabling direct usage of by-products containing FFAs and glycerol [131]. Since industrial quantities of FFAs can be obtained from other industrial production processes, esterification can be used to produce SLs for food applications. Currently, esterification reactions catalyzed by lipases are mainly used in the synthesis of DGs, such as 1,3-diolein [132]. In addition, the common method of producing SLs through an esterification-based process involves MLCTs. Yang et al. [133] implemented a one-step esterification process in the production of MLCTs based on caprylic acid, capric acid and oleic acid with glycerol, resulting in 72.19% MLCT yield. However, there are many other glyceride by-products of one-step esterification, such as DAGs and MAGs, and the fatty acids can only be randomly arranged on the glycerol backbone. Nagao et al. [134] used esterification for the synthesis of SL-containing DHA. The product contained 51.7 mol% DHA at sn-1,3 positions, while the sn-2 DHA content was 17.3 mol%. This can be attributed to the large steric hindrance of LC-PUFAs and the higher catalytic activity of lipase for other fatty acids.

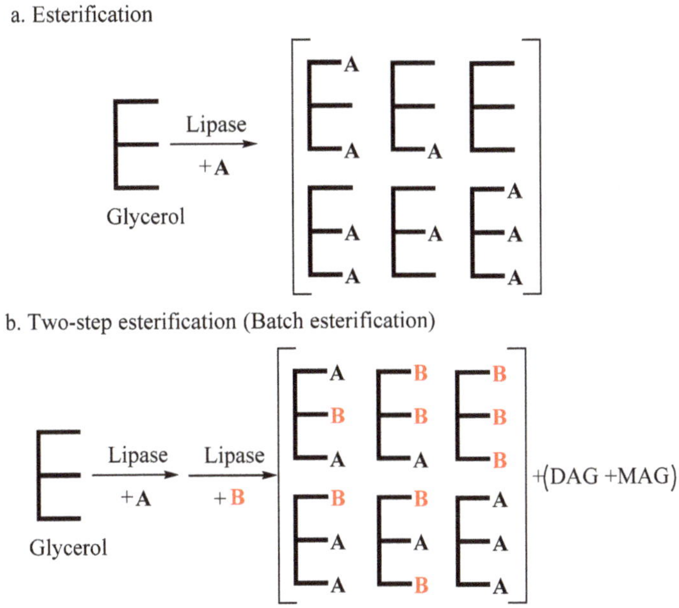

Figure 6. Process routes of **one-step (a) and two-step (b)** enzymatic esterification (A–B: fatty acid).

Two-step esterification has been used to synthesize DAGs and TAGs step by step, which includes adding FFAs in batches. Agapay et al. [135] used two-step esterification for the preparation of OPO, with oleic acid, palmitic acid and glycerol as reaction substrates. The OPO content in the prepared product was 34.98–39.55 wt%, which matched that of acidolysis products in several studies. In the study of Medina et al. [136], a 1,3-dicapryloyl-2-PUFA glycerol product containing about 90 wt% TAGs was synthesized using concentrated PUFAs from sardine discards through two-step esterification. The two-step process usually utilizes different lipases for each step and involves the intermediate purification of precursors between the steps.

3.2.4. Combination of Alcoholysis and Reesterification

In order to obtain SLs with high purity, the combination of alcoholysis and reesterification has been conducted by a large number of researchers. As is shown in Figure 7, the first step is alcoholysis between TAGs and ethanol under the catalysis of sn-1,3-specific lipase to produce sn-2 MAGs, which are not stable and are easy to convert to sn-1,3 MAGs. Then, the second step is acidolysis between MAGs and fatty acids.

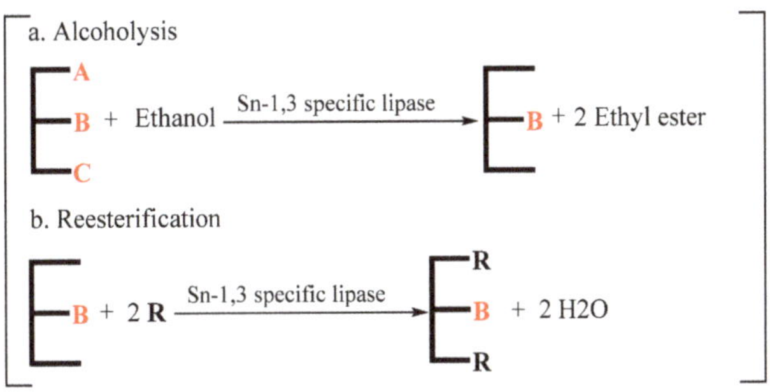

Figure 7. Process route of enzymatic alcoholysis (**a**) and reesterification (**b**) (A–C, R: fatty acid).

Since the product of this reaction retains fatty acids at the sn-2 position to the greatest extent, this method is often used to prepare SLs rich in specific fatty acids at the sn-2 position, such as sn-2 palmitate SL and SLs rich in sn-2 LC-PUFAs. Schmid et al. [137] and Pfeffer et al. [138] synthesized OPO through a two-step enzymatic process catalyzed by sn-1,3-specific lipases, which included alcoholysis of tripalmitic acid with ethanol to produce sn-2 palmitate MAG and then esterification between sn-2 MAGs and oleic acid. After purification, the obtained products contained 96% sn-2 palmitic acid, 90% sn-1,3 oleic acid and 95% OPO purity. Irimescu et al. [139] selected tridocosahexaenoylglycerol and ethylcaprylate as raw materials and obtained products (1,3-dicapryloyl-2-docosahexaenoyl glycerol (CDC), 85.4% yield) with regioisomeric purity reaching 100% after a two-step enzymatic process.

Table 5 summarizes the main studies on the preparation of SLs by enzymatic methods. Although the purity of SLs obtained by a two-step enzymatic method is very high, the complicated steps make this method less attractive.

3.2.5. Fermentation with Microorganisms

All of the enzymatic methods require expensive lipases, which limits their industrial feasibility [154]. Although immobilized lipase can be recovered, a decrease in catalytic activity is inevitable [155]. Therefore, more green and sustainable methods for the synthesis of HMFSs need to be developed. Microorganisms may be a good choice for OPO production. Microbial oils have many advantages and have been identified as a promising alternative to vegetable oils and animal fats because of their season-independent production and adjustable composition [156].

The potential for producing HMFSs by fermentation using *Rhodococcus opacus* was explored in the study of Zhang et al. [152]. Transcriptome analysis showed that β-oxidation, fatty acid elongation and Kennedy pathways exist in *R. opacus*. The fatty acid supplied as a carbon source can enter the Kennedy pathway directly or through a de novo biosynthetic pathway according to the chain length. As is shown in Figure 8, the results indicated that different compounds with chain lengths from 12 to 18, used as a carbon source, could be incorporated into TAGs directly. PUFAs, including ARA, EPA and DHA, could enter the Kennedy pathway directly and be involved in the biosynthesis of TAGs. Specifically, if the chain length was less than 12, the carbon source entered the fatty acid elongation pathway and formed palmitic acid or oleic acid. When oleic acid, palmitic acid and linoleic acid were supplied as substrate fats, 1-oleoyl-2-palmityl-3-linoleoyl glycerol (OPL) was formed [152,153].

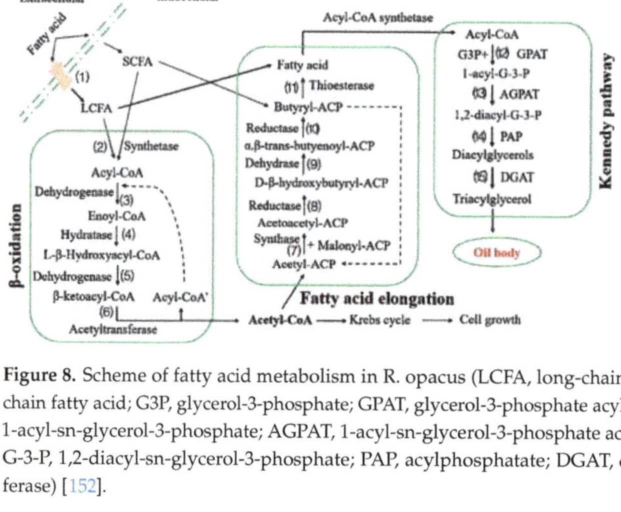

Figure 8. Scheme of fatty acid metabolism in R. opacus (LCFA, long-chain fatty acid; SCFA, short-chain fatty acid; G3P, glycerol-3-phosphate; GPAT, glycerol-3-phosphate acyltransferase; 1-acyl-G-3-P, 1-acyl-sn-glycerol-3-phosphate; AGPAT, 1-acyl-sn-glycerol-3-phosphate acyltransferase; 1,2-diacyl-G-3-P, 1,2-diacyl-sn-glycerol-3-phosphate; PAP, acylphosphatate; DGAT, diacylglycerol acyltransferase) [152].

Table 5. Studies of SL synthesis for infant formula.

Reference	Type of Reaction	Raw Materials	Solvent System	Lipase	Results
[129]	Acidolysis	Tripalmitin + caprylic acid	Hexane	Lipozyme® TL IM (*T. lanuginosus* lipase)	Caprylic acid incorporation = 44.9 mol%
[140]	Acidolysis	Tripalmitin + oleic acid	Solvent-free	*Heterologous Rhizopus oryzae* lipase	Oleic acid incorporation = 22–30 mol%
[141]	Acidolysis	Tripalmitin + oleic acid	(1) Hexane (2) Solvent-free	(1) Lipozyme® TL IM (2) Lipozyme® RM IM	(1) Sn-2 palmitic acid = 92.92% OPO = 32.34% (2) Sn-2 palmitic acid = 86.62% OPO = 40.23%
[142]	Acidolysis	(1) Tripalmitin + oleic acid (2) Lard + oleic acid (3) Restructured palm oil + oleic acid	Hexane	*Candida lipolytica* lipase	(1) OPO = 46.5% (2) OPO = 45.0% (3) OPO = 32.4%
[143]	Acidolysis	Hazelnut oil + FFAs	Hexane	Novozym® 435 (*C. antarctica fraction B* lipase)	n-3 PUFAs = 19.6%
[144]	Acidolysis	Microbial oil + oleic acid	Solvent-free	Lipozyme® RM IM	Sn-2 ARA = 49.71% Sn-1,3 oleic acid = 47.05%
[145]	Acidolysis	OMEGA-GOLD oil + capric acid	Solvent-free	PS-30 lipase	Sn-2 DHA = 27.9% Sn-2 DPA = 12.6% Sn-1,3 capric acid = 13.3%
[146]	Batch acidolysis	(1) TGA40 oil + caprylic acid (2) TGA58F oil + caprylic acid (3) TGA55E oil + caprylic acid	Solvent-free	Immobilized *R. oryzae* lipase	(1) 1, 3-Capryloyl-2-arachidonoyl glycerol (CAC) = 36.0 mol% (2) CAC = 43.1 mol% (3) CAC = 50.7 mol%
[147]	(1) Acidolysis (2) Two-step acidolysis	Fungal oil + tripalmitin + oleic acid	Solvent-free	Lipozyme® RM IM	(1) Oleic acid incorporation = 37.6% (2) Oleic acid incorporation = 55.4%
[122]	(1) Hydrolysis (2) Acidolysis	DHASCO + ARASCO + tripalmitin	Hexane	Lipozyme® TL IM	ARA = 17.69%; DHA = 10.75%; Sn-2 palmitic acid = 48.53%
[148]	(1) Acidolysis (2) (2)–(7): Inter-esterification	(1) Tripalmitin + oleic acid (2) Palm oil + olive oil (3) Palm oil + camellia oil (4) Palm oil + earth almond oil (5) Lard + olive oil (6) Lard + camellia oil (7) Lard + earth almond oil	Isooctane	Lipozyme® IM-20	(1) OPO = 55.2% (2) OPO = 21.8% (3) OPO = 25.2% (4) OPO = 19.9% (5) OPO = 12.9% (6) OPO = 15.4% (7) OPO = 9.0%
[149]	(1) Saponification (2) Fractionation (3) Esterification (4) Transesterification (5) Acidolysis	Oleic acid + refined palm oil	n-Hexane	Lipase IM 60	OPO = 74 mol%
[150]	Interesterification	Tripalmitin + ethyl oleate	Solvent-free	Lipozyme® TL IM	Sn-2 palmitic acid = 80.6% Sn-1,3 oleic acid = 64.9%
[130]	Two-step interesterification	Amaranth oil + ethyl palmitate + DHASCO	Solvent-free	(1) Novozym® 435 (2) Lipozyme® RM IM	palmitic acid=33.9%; DHA = 1.9%
[134]	Esterification	DHA(by-product) + glycerol	Solvent-free	Immobilized *R. miehei* lipase	Sn-2 DHA = 17.3 mol% Sn-1,3 DHA = 51.7 mol%
[135]	Two-step esterification	Oleic acid + palmitic acid + glycerol	Solvent-free	Novozym® 435	OPO = 34.98–39.55 wt.%
[139]	(1) Alcoholysis (2) Reesterification	DDD/EEE + ethanol + ethylcaprylate	(1) Ethanol (2) Solvent-free	(1) Novozym® 435 (2) Lipozyme® IM	1. 1,3-Dicapryloyl-2-docosahexaenoyl glycerol yield = 85.4% Regioisomeric purity = 100% 2. 1,3-Dicapryloyl-2-eicosapentaenoyl glycerol yield = 84.2% Regioisomeric purity = 99.8%
[151]	(1) Alcoholysis (2) Reesterification	Bonito oil +ethanol + caprylic acid	(1) Ethanol (2) Solvent-free	Novozym® 435	(1) Sn-2 DHA MAG = 43.5% (2) CA-C22-CA = 51 wt.%
[137]	(1) Alcoholysis (2) Esterification	Tripalmitin + ethanol + oleic acid	(1) Organic solvent (2) Solvent-free	Immobilized lipase from *Rhizomucor miehei* and *Rhizopus delemar*	OPO yield = 78%
[138]	(1) Alcoholysis (2) Esterification	Tripalmitin + ethanol + oleic acid	(1) Acetone (2) Hexane	Novozym® 435	OPO purity = 95% OPO yield = 90%
(1) [152] (2) [153]	Fermentation	Chemically interesterified fat/ mixture of ethyl oleate/ ethyl palmitate	-	*Rhodococcus opacus*	(1) OPL = 40.09% (2) OPO = 47.1%

In the study of Zhang et al. [153], chemically interesterified fat or a mixture of ethyl oleate/ethyl palmitate 2:1 (*w*/*w*) was used as a starting material. After fermentation with *R. opacus*, OPO (47.1%), OPL (13.9%), PPO (13.1%) and PPoO (16:0–16:1–18:1) (10.3%) were the most abundant TAG species. Its feasibility of large-scale production has also been proven. The main component of the oil produced by *R. opacus* is TAG, which shows that it can be used as edible oil after proper refining. Furthermore, compared with the enzymatic method, no acyl group migration was detected during the fermentation process [152].

3.3. *Purification and Treatment of SLs*

Regardless of whether HMFS is prepared by chemical or enzymatic catalysis, the reaction products contain by-products, such as FFAs and DAGs. Removing by-products by effective methods improves not only the quality but also the nutritive value of HMFSs. Purification methods mainly include physical methods and chemical methods [157,158]. Chemical purification methods, such as urea inclusion and solvent crystallization, consume a large amount of solvent that is difficult to fully recover, resulting in high cost and solvent residue [159,160]. Therefore, chemical methods are often used in laboratories. Physical methods include supercritical carbon dioxide extraction, distillation, molecular distillation, etc. Molecular distillation is a safe, green and rapid purification method. It has the advantages of a high degree of separation, short material residence time and low distillation temperature. It is suitable for the separation of substances with a high boiling point and high heat sensitivity [115,150].

From a stability point of view, PUFAs are highly unstable and tend to oxidize. PUFAs have at least two double bonds, which are easily attacked by active agents. The resulting PUFA hydroperoxides may lead to secondary reactions, a bad smell and bioactivity loss [161]. Therefore, extra protection for PUFA products is necessary. LC-PUFAs in infant formula are mostly added through the microcapsule process. In the microcapsule process, natural or synthetic polymer materials are used as capsule materials, and SLs are embedded in wall materials by chemical, physical or physicochemical methods [162]. This process can protect LC-PUFA-containing SLs, mask the bad flavor and release SLs at a controllable rate under specific conditions. There are many methods to achieve microencapsulation, such as spray condensation, freeze-drying, etc. [163,164]. Among them, spray drying is the most widely used method in the food industry [165]. The principle is that the emulsion is atomized, and the solvent is rapidly evaporated by high-temperature airflow. Finally, the wall material forms a membrane structure on the core surface. The process has several advantages, including low cost, high controllability of process conditions, uniform particles and good solubility of the product [166]. For the microencapsulation of LC-PUFAs, the wall material should be food-grade ingredients and be suitable for infants, among which natural ingredients are preferred. Milk protein, a milk protein–carbohydrate mixture and their Maillard products are good wall materials [167]. In order to prevent the oxidation of LC-PUFAs during microencapsulation, some natural antioxidant components, such as tocopherol, can be added. This can further improve the stability of LC-PUFAs in infant formula [168].

3.4. *Preparation of MFGM*

PLs have been added to infant formula as biologically active elements due to their potential benefits for the optimal growth and health of infants. MFGM supplements and polar lipid-rich ingredients are mainly extracted from bovine milk. As is shown in Figure 9, the separation method of the MFGM can generally be divided into four steps: fat globule separation, cream washing, release of the MFGM from the globules and collection of MFGM materials [169]. Membrane proteins are highly vulnerable and easily lost during isolation, especially loosely bound proteins; around 16% of MFGM proteins were lost during production. Luckily, only 4% of phospholipids were lost [170]. Washing also causes losses of tocopherol, which is an antioxidant in the MFGM [171]. Bovine serum albumin could still be detected in the MFGM product despite successive washings [172].

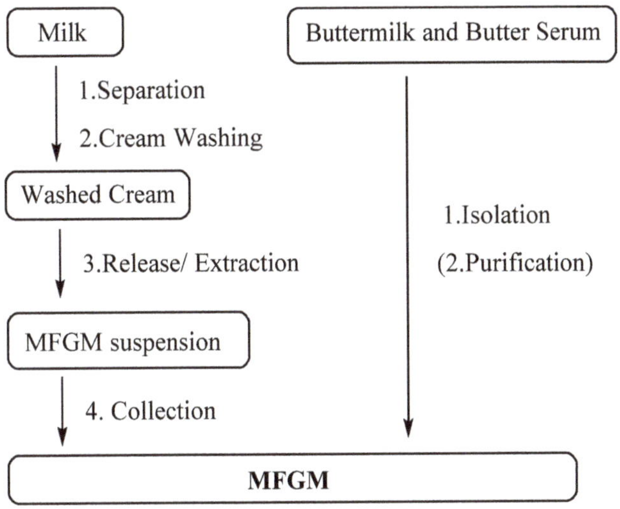

Figure 9. Process route of MFGM production.

The preparation of the MFGM from fresh industrial liquid butter serum or buttermilk mainly requires isolation, including micro- and ultrafiltration, and renneting coagulation. Citrate has also been used to dissociate casein micelles [173–175]. More phospholipids can be separated by further purification, including supercritical fluid extraction and solvent extraction.

3.5. Commercial HMFS Products

Due to the positive effect on infant health, the most popular commercial HMFS product is sn-2 palmitate SL [176]. A variety of sn-2 palmitate SL products on the market and their compositions are listed in Table 6. As the first commercialized sn-2 palmitate SL product developed by IOI Loders Croklaan, Betapol® has been used in infant formulas since 1995 [177]. The main components in Betapol® are OPO and 1-oleoyl-2-palmitoyl-3-linoleoylglycerol/1-linoleyl-2-palmitoyl-3-oleoylglycerol (OPL/LPO). In addition, Advanced Lipids developed InFat®, in which 70–75% of palmitic acid is distributed at the sn-2 position [178,179]. In order to meet the needs of the Chinese market, these companies have developed OPO products that mimic the composition of Chinese HMF, such as Betapol® Select and InFat® Plus. In addition, Wilmar developed Milkopas® 9320, in which over 62% of palmitic acid is distributed at the sn-2 position, and the content of OPO is over 53%. The contents of sn-2 palmitic acid and OPO are higher than the Chinese national standard and much closer to the contents in HMF. As one step closer to the goal of "humanization", Milkopas® has taken one-third of the current market share. As an LC-PUFA product, Nature's One developed Baby's Only® Essentials DHA & ARA Fatty Acid, which is naturally derived from egg phospholipids using an aqueous process [107].

Table 6. Main composition of commercial OPO products.

Composition *	IOI Loders Croklaan		Advanced Lipids		Wilmar	
	Betapol®	Betapol® Plus	InFat®	InFat® Plus	Milkopas® 9100	Milkopas® 9320
Sn-2 PA (%)	55	65–75	52	52-	>52	>62
OPO (TG 52:2, %)	40–45	-	56–60	35–40	>40	>53
OPL (TG 53:2, %)	-	-	24	42	-	-

* PA: palmitic acid; OPO: 1,3-oleoyl-2-palmitoylglycerol; OPL: 1-oleoyl-2-palmitoyl-3-linoleoylglycerol.

Some MFGM supplement products are already on the market and have been added to some infant formulas in the Chinese market. For instance, Lacprodan® MFGM-10 by Arla Foods Ingredients Group is a protein-rich MFGM supplement derived from bovine milk, containing 73% protein, 8% MFGM protein, 7% phospholipid, 2% sialic acid, 5% IgG, 0.2% lactoferrin and 0.2% ganglioside [180], and BSC by Fonterra Co-operative Group contains 52.30 g protein, 36.20 g lipid with 13.67 g phospholipid and 0.63 g ganglioside, 6.60 g lactose and 5.20 g minerals per 100 g. The MFGM from Fonterra consists of 647 mg/L phospholipids, including 57 mg/L PI, 80 mg/L PS, 173 mg/L PC and 141 mg/L SM. In addition, Inpulse from Büllinger SA has a nine-fold higher concentration of phospholipids compared to that found in Lacprodan® PL-20 (Arla) and contains 16% phospholipid (SM, gangliosides) and about 55% proteins [181].

4. Fat-Related Regulations and Guidelines for Infant Formula

In 1938, the U.S. Federal Food, Drug, and Cosmetic Act (FFDCA) contained the first reference to foods for special dietary purposes, including infant formulas [182]. The act helped standardize the branding of food, setting standards for identity, quality and fill volume or weight [183]. The Infant Formula Act of 1980 clarified the US Food and Drug Administration (FDA)'s authority to establish minimum nutrient requirements and to establish quality control procedures [184]. It also established maximum permissible levels of protein, fat, sodium, potassium, chloride and vitamins A and D for the first time.

At present, the current legislation on infant formula is set by different authorities, including the Codex Alimentarius Commission (CAC), FDA, the European Commission (EC) and the National Health Commission of the People's Republic of China (NHC) [107]. The maximum and the minimum contents legislated for infant formula lipids are summarized in Table 7. As the compulsory composition, total fat and linoleic acid content within 0.3–3.0 g/100 kcal are required by all authorities. ALA is required to be more than 50 mg/100 kcal by CAC, EC and NHC. Trans fatty acids should be less than 3% of total fatty acid, and erucic acid should be less than 1% of total fatty acid. Phospholipids should not exceed 2 g/L, according to CAC and EC. LC-PUFAs are permitted as facultative additions by CAC, EC and NHC [185,186].

Table 7. Legislated contents of fat in infant formula [185,186].

Component	Min	Max
Total fat (g/100 kcal)	3.3	6.5
Linoleic acid (g/100 kcal)	0.3	3.0
Alinoleic acid (mg/100 kcal)	50.0	-
Trans fatty acid (%)	-	3.0
Erucic acid (%)	1.0	2.0
ARA (%)	-	1.0
DHA (%)	-	2.0
Phospholipids (g/L)	-	2.0
Choline (mg/100 kcal)	7.0	50.0

DHA is now recommended as a required constituent. In 2005, the Nutrition Committee of the European Society of Pediatric Gastroenterology, Hepatology and Nutrition (ESPGHAN) proposed a global standard for the composition of infant formula. The standard includes 7 and 50 mg choline/100 kcal as the recommended minimum and maximum values, respectively [187]. Requirements for infant formula in the USA indicate a minimum of 7 mg/100 kcal for infant formula in general [188–191]. The Department of National Health and Welfare Canada established that the choline content of formula should be no less than 12 mg/100 kcal. The Food and Nutrition Board [192] has set an adequate intake of choline of 125 mg/day for infants from 0 to 6 months of age.

5. Evaluation of HMFSs

With the popularization of infant formula, a wide variety of commercial products for infant formula with varying quality emerged. Meanwhile, more and more studies have been working on the analysis and evaluation of the composition of HMFS [92,193,194]. In order to further evaluate the quality and potential application prospects of synthetic HMFSs, it is also vital to evaluate the quality of refined HMFS products. The characterization of the conventional physical and chemical indicators of the product include saponification value, acid value, peroxide value and oxidation stability [195–197].

Due to its advantage in the absorption of fatty acids, minerals and vitamins and its ability to reduce constipation, HMF is regarded as the "golden standard" and provides a direction for the development of HMFSs [5–7]. Therefore, the evaluation of the similarity between HMFSs and HMF is also indispensable. However, the wide variety and the complex positional distribution of fatty acids in HMF have made the evaluation difficult. Although composition analysis can clarify the difference in compositions between HMFSs and HMF, it cannot intuitively and quantitatively evaluate the similarity [8,198]. In order to evaluate the similarity of HMFSs, it is necessary to establish scoring models. A scoring model will promote the development of the infant formula industry towards standardization. At present, there are only a few reports on establishing similarity evaluation methods of HMFSs.

Wang proposed a two-dimensional scoring model based on the principle of "deduction"; that is, the total and sn-3 fatty acids in HMF are used as a reference standard [8]. The greater the deviation between the corresponding content of HMFSs and the reference value of HMF, the lower the similarity score. The effectiveness of the evaluation model was verified by evaluating 15 different fats in this study, of which 5 samples were infant formula milk powder. The results showed that lipids in infant formula had a high similarity to HMF in terms of fatty acid composition but a generally low similarity in terms of the distribution of fatty acids [8].

In addition, Zou et al. [198] proposed a four-dimensional scoring model based on the main fatty acids, the relative content of sn-2 fatty acids, PUFAs and the main TAGs of HMF as reference standards. The authors evaluated three infant formula samples using this model, and the results showed that the similarity of the main fatty acids was relatively high (89.4–98.4), but the similarity of the relative content of sn-2 fatty acids and PUFAs was generally low, with values of 40.2–58.6 and 28.9–67.2, respectively [198].

In a recent study, Kloek et al. [199] extended the model of Wang et al. [8] and included sn-1,3 fatty acid composition, and fatty acids at sn-1,3 positions were weighted two-thirds in the average index. Fatty acids positioned at sn-1,3 could be linked to fatty acid soap formation and calcium bioavailability in infants. Therefore, sn-1,3 fatty acids should be involved in the evaluation of HMFSs. According to their results, a higher average SI was observed in bovine milk fat–containing infant formulas compared with palm oil–containing or palm oil–free infant formulas [199].

The first two classic scoring models have been applied for evaluations by researchers [200–204]. Zou et al. [205] evaluated the potential of mammalian milk fats for HMFSs based on the total and sn-2 fatty acid, TAG, phospholipid and phospholipid fatty acid compositions. According to the results, all of the milk fats tested had high degrees of similarity to HMF in total fatty acid composition. However, the degrees of similarity in other chemical aspects were low, indicating that these milk fats did not meet the requirements for HMFSs [205]. Zou et al. [200] also used a scoring model to evaluate the similarity of HMFS synthesized by an enzymatic process, in which the relative content of sn-2 fatty acids and the similarity of TAGs were improved. In addition, the lipids in commercial infant formulas were evaluated by the models. The results indicated that the scores of vegetable oil–based formulas were in the range of 0–49.5, whereas the scores of formulas based on cow and goat milk were in the ranges of 5.7–53.2 and 26.4–52.7, respectively. All commercial HMFSs had scores below 60. The overall score of 180 commercial HMFSs was 38.4 [100,204,206].

6. Conclusions

Human milk is the first-choice food for infants, and HMF plays an important role in the development of infants. In the past few years, great progress has been made in HMFS preparation. At present, HMFSs can effectively simulate the fatty acid composition of HMF, and LC-PUFAs are supplied through the addition of deep-sea fish oil or microalgae oil to meet the needs of infants. With the in-depth study of HMF, its composition and structure are well known. More attention is paid to the more in-depth simulation of HMFSs at the TAG level. Thus, the first challenge in the development of HMFSs is the evaluation of HMFSs. The scoring model should be further studied at the TAG level or even in the aspect of the milk fat globule. This could further inform the research and production of HFMSs. At present, the synthesis of SLs, such as sn-2 palmitate SL and LC-PUFA-enriched SLs, is mainly synthesized by acidolysis and the combination of alcoholysis and esterification. However, the high cost of enzymatic synthesis limits the application of SLs. The production cost should be further reduced through research for more suitable food-grade lipase, immobilization materials, substrates and process methods to realize industrial production of SLs. In addition, more attention should be paid to fermentation synthesis and the formulation of relevant laws and regulations. Fermentation offers a green and sustainable option and has the potential application in industrial production. In the aspect of the milk globule, the HMFSs in infant formula still have some deficiencies compared to HMF, affecting the digestion efficiency of the lipid. With the full awareness of knowledge limitations, we hope to encourage additional research on HMFSs.

Author Contributions: Conceptualization, X.J. and X.Z.; formal analysis, Z.C.; investigation, X.J.; data curation, X.X.; writing—original draft preparation, X.J.; writing—review and editing, X.Z.; visualization, Z.C.; supervision, X.Z. All authors have read and agreed to the published version of the manuscript.

Funding: This research was funded by Province Key Laboratory of Cereal Resource Transformation and Utilization, Henan University of Technology, grant number PL2020002, and the National Natural Science Foundation of China, grant number 31601433.

Institutional Review Board Statement: Not applicable.

Informed Consent Statement: Not applicable.

Data Availability Statement: Not applicable.

Conflicts of Interest: The authors declare no conflict of interest.

References

1. Eriksen, K.G.; Christensen, S.H.; Lind, M.V.; Michaelsen, K.F. Human milk composition and infant growth. *Curr. Opin. Clin. Nutr. Metab. Care* **2018**, *21*, 200–206. [CrossRef] [PubMed]
2. Mosca, F.; Giannì, M.L. Human milk: Composition and health benefits. *Med. Surg. Pediatr.* **2017**, *39*, 47–52. [CrossRef] [PubMed]
3. Saarela, T.; Kokkonen, J.; Koivisto, M. Macronutrient and energy contents of human milk fractions during the first six months of lactation. *Acta Paediatr.* **2005**, *94*, 1176–1181. [CrossRef]
4. Arab-Tehrany, E.; Jacquot, M.; Gaiani, C.; Imran, M.; Desobry, S.; Linder, M. Beneficial effects and oxidative stability of omega-3 long-chain polyunsaturated fatty acids. *Trends Food Sci. Technol.* **2012**, *25*, 24–33. [CrossRef]
5. Hamosh, M.; Bitman, J.; Wood, D.L.; Hamosh, P.; Mehta, N. Lipids in milk and the first steps in their digestion. *Pediatrics* **1985**, *75*, 146–150. [CrossRef] [PubMed]
6. Jensen, R.G. Lipids in human milk. *Lipids* **1999**, *34*, 1243–1271. [CrossRef] [PubMed]
7. Boersma, E.R.; Offringa, P.J.; Muskiet, F.A.; Chase, W.M.; Simmons, I.J. Vitamin E, lipid fractions, and fatty acid composition of colostrum, transitional milk, and mature milk: An international comparative study. *Am. J. Clin. Nutr.* **1991**, *53*, 1197–1204. [CrossRef]
8. Wang, Y.-H.; Mai, Q.-Y.; Qin, X.-L.; Yang, B.; Wang, Z.-L.; Chen, H.-T. Establishment of an evaluation model for human milk fat substitutes. *J. Agric. Food Chem.* **2010**, *58*, 642–649. [CrossRef]
9. Keenan, T.W.; Patton, S. The structure of milk: Implications for sampling and storage: A. The milk lipid globule membrane. In *Handbook of Milk Composition*; Elsevier: Amsterdam, The Netherlands, 1995; pp. 5–50.
10. Michalski, M.-C.; Briard, V.; Michel, F.; Tasson, F.; Poulain, P. Size distribution of fat globules in human colostrum, breast milk, and infant formula. *J. Dairy Sci.* **2005**, *88*, 1927–1940. [CrossRef]

11. Lopez, C.; Ménard, O. Human milk fat globules: Polar lipid composition and in situ structural investigations revealing the heterogeneous distribution of proteins and the lateral segregation of sphingomyelin in the biological membrane. *Colloids Surf. B* **2011**, *83*, 29–41. [CrossRef]
12. Lopez-Lopez, A.; Lopez-Sabater, M.; Campoy-Folgoso, C.; Rivero-Urgell, M.; Castellote-Bargallo, A. Fatty acid and sn-2 fatty acid composition in human milk from Granada (Spain) and in infant formulas. *Eur. J. Clin. Nutr.* **2002**, *56*, 1242–1254. [CrossRef] [PubMed]
13. Straarup, E.M.; Lauritzen, L.; Faerk, J.; Høy, C.-E.; Michaelsen, K.F. The stereospecific triacylglycerol structures and fatty acid profiles of human milk and infant formulas. *J. Pediatr. Gastroenterol. Nutr.* **2006**, *42*, 293–299. [CrossRef] [PubMed]
14. Innis, S.M.; Dyer, R.; Nelson, C.M. Evidence that palmitic acid is absorbed as sn-2 monoacylglycerol from human milk by breast-fed infants. *Lipids* **1994**, *29*, 541–545. [CrossRef] [PubMed]
15. Carnielli, V.P.; Luijendijk, I.H.; Van Goudoever, J.B.; Sulkers, E.J.; Boerlage, A.A.; Degenhart, H.J.; Sauer, P.J. Structural position and amount of palmitic acid in infant formulas: Effects on fat, fatty acid, and mineral balance. *J. Pediatr. Gastroenterol. Nutr.* **1996**, *23*, 553–560. [CrossRef]
16. Jensen, R.G.; Ferris, A.M.; Lammi-Keefe, C.J.; Henderson, R.A. Lipids of bovine and human milks: A comparison. *J. Dairy Sci.* **1990**, *73*, 223–240. [CrossRef]
17. Francois, C.A.; Connor, S.L.; Wander, R.C.; Connor, W.E. Acute effects of dietary fatty acids on the fatty acids of human milk. *Am. J. Clin. Nutr.* **1998**, *67*, 301–308. [CrossRef]
18. Luukkainen, P.; Salo, M.K.; Nikkari, T. Changes in the fatty acid composition of preterm and term human milk from 1 week to 6 months of lactation. *J. Pediatr. Gastroenterol. Nutr.* **1994**, *18*, 355–360. [CrossRef]
19. Thakkar, S.K.; De Castro, C.A.; Beauport, L.; Tolsa, J.-F.; Fischer Fumeaux, C.J.; Affolter, M.; Giuffrida, F. Temporal progression of fatty acids in preterm and term human milk of mothers from Switzerland. *Nutrients* **2019**, *11*, 112. [CrossRef]
20. Koletzko, B.; Rodriguez-Palmero, M.; Demmelmair, H.; Fidler, N.; Jensen, R.; Sauerwald, T. Physiological aspects of human milk lipids. *Early Hum. Dev.* **2001**, *65*, S3–S18. [CrossRef]
21. Tinoco, S.M.B.; Sichieri, R.; Setta, C.L.; Moura, A.S.; do Carmo, M.d.G.T. Trans fatty acids from milk of Brazilian mothers of premature infants. *J. Paediatr. Child Health* **2008**, *44*, 50–56. [CrossRef]
22. Moltó-Puigmartí, C.; Castellote, A.I.; Carbonell-Estrany, X.; López-Sabater, M.C. Differences in fat content and fatty acid proportions among colostrum, transitional, and mature milk from women delivering very preterm, preterm, and term infants. *Clin. Nutr.* **2011**, *30*, 116–123. [CrossRef] [PubMed]
23. Bobiński, R.; Bobińska, J. Fatty acids of human milk—A review. *Int. J. Vitam. Nutr. Res.* **2020**, *21*, 1–12. [CrossRef] [PubMed]
24. Ni, M.; Wang, Y.; Wu, R.; Zhang, L.; Xu, X.; Yang, Y.; Chen, J. Total and Sn-2 Fatty Acid Profile in Human Colostrum and Mature Breast Milk of Women Living in Inland and Coastal Areas of China. *Ann. Nutr. Metab.* **2021**, *77*, 29–37. [CrossRef] [PubMed]
25. De la Presa-Owens, S.; López-Sabater, M.; Rivero-Urgell, M. Fatty acid composition of human milk in Spain. *J. Pediatr. Gastroenterol. Nutr.* **1996**, *22*, 180–185. [CrossRef] [PubMed]
26. Koletzko, B.; Thiel, I.; Abiodun, P.O. The fatty acid composition of human milk in Europe and Africa. *J. Pediatr.* **1992**, *120*, S62–S70. [CrossRef]
27. Hinton, A., Jr.; Ingram, K.D. Antimicrobial activity of potassium hydroxide and lauric acid against microorganisms associated with poultry processing. *J. Food Prot.* **2006**, *69*, 1611–1615. [CrossRef]
28. Qi, C.; Sun, J.; Xia, Y.; Yu, R.; Wei, W.; Xiang, J.; Jin, Q.; Xiao, H.; Wang, X. Fatty acid profile and the sn-2 position distribution in triacylglycerols of breast milk during different lactation stages. *J. Agric. Food Chem.* **2018**, *66*, 3118–3126. [CrossRef]
29. Jensen, R.G. The lipids in human milk. *Prog. Lipid Res.* **1996**, *35*, 53–92. [CrossRef]
30. Innis, S.M. Dietary triacylglycerol structure and its role in infant nutrition. *Adv. Nutr.* **2011**, *2*, 275–283. [CrossRef]
31. Abedi, E.; Sahari, M.A. Long-chain polyunsaturated fatty acid sources and evaluation of their nutritional and functional properties. *Food Sci. Nutr.* **2014**, *2*, 443–463. [CrossRef]
32. Qi, K.; Hall, M.; Deckelbaum, R.J. Long-chain polyunsaturated fatty acid accretion in brain. *Curr. Opin. Clin Nutr Metab. Care* **2002**, *5*, 133–138. [CrossRef] [PubMed]
33. McCann, J.C.; Ames, B.N. Is docosahexaenoic acid, an n−3 long-chain polyunsaturated fatty acid, required for development of normal brain function? An overview of evidence from cognitive and behavioral tests in humans and animals. *Am. J. Clin. Nutr.* **2005**, *82*, 281–295. [CrossRef] [PubMed]
34. Makrides, M.; Gibson, R.A. Long-chain polyunsaturated fatty acid requirements during pregnancy and lactation. *Am. J. Clin. Nutr.* **2000**, *71*, 307–311. [CrossRef] [PubMed]
35. Koletzko, B.; Agostoni, C.; Carlson, S.E.; Clandinin, T.; Hornstra, G.; Neuringer, M.; Uauy, R.; Yamashiro, Y.; Willatts, P. Long chain polyunsaturated fatty acids (LC-PUFA) and perinatal development. *Acta Paediatr.* **2001**, *90*, 460–464. [CrossRef]
36. Das, U. Long-chain polyunsaturated fatty acids in the growth and development of the brain and memory. *Nutrition* **2003**, *19*, 62. [CrossRef]
37. Gottrand, F. Long-chain polyunsaturated fatty acids influence the immune system of infants. *J. Nutr.* **2008**, *138*, 1807–1812. [CrossRef]
38. Miles, E.A.; Childs, C.E.; Calder, P.C. Long-chain polyunsaturated fatty acids (LCPUFAs) and the developing immune system: A narrative review. *Nutrients* **2021**, *13*, 247. [CrossRef]

39. Booyens, J.; Engelbrecht, P.; Roux, S.L.; Louwrens, C.C.; Merwe, C.V.; Katzeff, I.E. Some effects of the essential fatty acids linoleic acid and alpha-linolenic acid and of their metabolites gamma-linolenic acid, arachidonic acid, eicosapentaenoic acid, docosahexaenoic acid, and of prostaglandins A1 and E1 on the proliferation of human osteogenic sarcoma cells in culture. *Prostaglandins Leukot. Med.* **1984**, *15*, 15–33.
40. Garcia, C.; Lutz, N.W.; Confort-Gouny, S.; Cozzone, P.J.; Armand, M.; Bernard, M. Phospholipid fingerprints of milk from different mammalians determined by 31P NMR: Towards specific interest in human health. *Food Chem.* **2012**, *135*, 1777–1783. [CrossRef]
41. Lopez, C. Lipid domains in the milk fat globule membrane: Specific role of sphingomyelin. *Lipid Technol.* **2010**, *22*, 175–178. [CrossRef]
42. Cilla, A.; Diego Quintaes, K.; Barberá, R.; Alegría, A. Phospholipids in human milk and infant formulas: Benefits and needs for correct infant nutrition. *Crit. Rev. Food Sci. Nutr.* **2016**, *56*, 1880–1892. [CrossRef] [PubMed]
43. Zeisel, S.H.; Blusztajn, J.K. Choline and human nutrition. *Annu. Rev. Nutr.* **1994**, *14*, 269–296. [CrossRef] [PubMed]
44. Zeisel, S.H.; Da Costa, K.A.; Franklin, P.D.; Alexander, E.A.; Lamont, J.T.; Sheard, N.F.; Beiser, A. Choline: An essential nutrient for humans. *FASEB J.* **1991**, *5*, 2093–2098. [CrossRef] [PubMed]
45. Zeisel, S.H. Choline: Critical role during fetal development and dietary requirements in adults. *Annu. Rev. Nutr.* **2006**, *26*, 229–250. [CrossRef]
46. Norris, G.H.; Milard, M.; Michalski, M.-C.; Blesso, C.N. Protective properties of milk sphingomyelin against dysfunctional lipid metabolism, gut dysbiosis, and inflammation. *J. Nutr. Biochem.* **2019**, *73*, 108224. [CrossRef]
47. Motouri, M.; Matsuyama, H.; Yamamura, J.-I.; Tanaka, M.; Aoe, S.; Iwanaga, T.; Kawakami, H. Milk sphingomyelin accelerates enzymatic and morphological maturation of the intestine in artificially reared rats. *J. Pediatr. Gastroenterol. Nutr.* **2003**, *36*, 241–247. [CrossRef]
48. Hannun, Y.A. The sphingomyelin cycle and the second messenger function of ceramide. *J. Biol. Chem.* **1994**, *269*, 3125–3128. [CrossRef]
49. Jensen, R.G. The composition of bovine milk lipids: January 1995 to December 2000. *J. Dairy Sci.* **2002**, *85*, 295–350. [CrossRef]
50. Heid, H.W.; Keenan, T.W. Intracellular origin and secretion of milk fat globules. *Eur. J. Cell Biol.* **2005**, *84*, 245–258. [CrossRef]
51. Jensen, R.G.; Newburg, D.S. Bovine milk lipids. In *Handbook of Milk Composition*; Elsevier: Amsterdam, The Netherlands, 1995; pp. 543–575.
52. Bode, L.; Beermann, C.; Mank, M.; Kohn, G.; Boehm, G.N. Human and bovine milk gangliosides differ in their fatty acid composition. *J. Nutr.* **2004**, *134*, 3016–3020. [CrossRef]
53. Idota, T.; Kawakami, H. Inhibitory effects of milk gangliosides on the adhesion of Escherichia coli to human intestinal carcinoma cells. *Biosci. Biotechnol. Biochem.* **1995**, *59*, 69–72. [CrossRef] [PubMed]
54. Rueda, R.; Sabatel, J.L.; Maldonado, J.; Molina-Font, J.A.; Gil, A. Addition of gangliosides to an adapted milk formula modifies levels of fecal Escherichia coli in preterm newborn infants. *J. Pediatr.* **1998**, *133*, 90–94. [CrossRef]
55. Abrahamse, E.; Minekus, M.; van Aken, G.A.; van de Heijning, B.; Knol, J.; Bartke, N.; Oozeer, R.; van der Beek, E.M.; Ludwig, T. Development of the Digestive System-Experimental Challenges and Approaches of Infant Lipid Digestion. *Food Dig.* **2012**, *3*, 63–77. [CrossRef] [PubMed]
56. Bourlieu, C.; Menard, O.; Alix, D.; Sams, L.; Rousseau, F.; Madec, M.N.; Robert, B.; Deglaire, A.; Pezennec, S.; Bouhallab, S. The structure of infant formulas impacts their lipolysis, proteolysis and disintegration during in vitro gastric digestion. *Food Chem.* **2015**, *182*, 224–235. [CrossRef]
57. Lien, E.L. The role of fatty acid composition and positional distribution in fat absorption in infants. *J. Pediatr.* **1994**, *125*, S62–S68. [CrossRef]
58. Mu, H.; Porsgaard, T. The metabolism of structured triacylglycerols. *Prog. Lipid Res.* **2005**, *44*, 430–448. [CrossRef]
59. Guo, Q.; Ye, A.; Bellissimo, N.; Singh, H.; Rousseau, D. Modulating fat digestion through food structure design. *Prog. Lipid Res.* **2017**, *68*, 109–118. [CrossRef]
60. Zhang, R.; Zhang, Z.; Zhang, H.; Decker, E.A.; McClements, D.J. Influence of lipid type on gastrointestinal fate of oil-in-water emulsions: In vitro digestion study. *Food Res. Int.* **2015**, *75*, 71–78. [CrossRef]
61. Shen, L.; Huang, W.; Xu, X.; Wang, L.; Wang, Q.; Li, S.; Yuan, X. Stool Saponified Fatty Acid, Behavior, Growth, and Stool Characteristics in Infants Fed a High-OPO Formula: A Randomized, Double-Blind Clinical Trial. *Front. Pediatr.* **2021**, *9*, 712201. [CrossRef]
62. He, X.; McClorry, S.; Hernell, O.; Lonnerdal, B.; Slupsky, C.M. Digestion of human milk fat in healthy infants. *Nutr. Res.* **2020**, *83*, 15–29. [CrossRef]
63. Filer, L., Jr.; Mattson, F.; Fomon, S. Triglyceride configuration and fat absorption by the human infant. *J. Nutr.* **1969**, *99*, 293–298. [CrossRef] [PubMed]
64. Brooke, O. Absorption of lard by infants. *Hum. Nutr. Appl. Nutr.* **1985**, *39*, 221–223. [PubMed]
65. Wan, J.; Hu, S.; Ni, K.; Chang, G.; Sun, X.; Yu, L. Characterisation of fecal soap fatty acids, calcium contents, bacterial community and short-chain fatty acids in sprague dawley rats fed with different sn-2 palmitic triacylglycerols diets. *PLoS ONE* **2016**, *11*, e0164894. [CrossRef] [PubMed]
66. Zhang, N.; Zeng, J.-P.; Wu, Y.-P.; Wei, M.; Zhang, H.; Zheng, L.; Deng, Z.-Y.; Li, J. Human Milk sn-2 Palmitate Triglyceride Rich in Linoleic Acid Had Lower Digestibility but Higher Absorptivity Compared with the sn-2 Palmitate Triglyceride Rich in Oleic Acid in Vitro. *J. Agric. Food Chem.* **2021**, *69*, 9137–9146. [CrossRef]

67. Wu, Y.; Zhang, N.; Deng, Z.-Y.; Zhang, H.; Li, J. Effects of the Major Structured Triacylglycerols in Human Milk on Lipid Metabolism of Hepatocyte Cells in Vitro. *J. Agric. Food Chem.* **2020**, *69*, 9147–9156. [CrossRef]
68. Whyte, R.; Whelan, D.; Hill, R.; McClorry, S. Excretion of dicarboxylic and ω-1 hydroxy fatty acids by low birth weight infants fed with medium-chain triglycerides. *Pediatr. Res.* **1986**, *20*, 122–125. [CrossRef]
69. Yuan, T.; Zhang, H.; Wang, X.; Yu, R.; Zhou, Q.; Wei, W.; Wang, X.; Jin, Q. Triacylglycerol containing medium-chain fatty acids (MCFA-TAG): The gap between human milk and infant formulas. *Int. Dairy J.* **2019**, *99*, 104545. [CrossRef]
70. Lee, K.T.; Akoh, C.C. Structured lipids: Synthesis and applications. *Food Rev. Int.* **1998**, *14*, 17–34. [CrossRef]
71. Lee, Y.-Y.; Tang, T.-K.; Phuah, E.-T.; Karim, N.A.A.; Alitheen, N.B.M.; Tan, C.-P.; Razak, I.S.A.; Lai, O.-M. Structural difference of palm based Medium-and Long-Chain Triacylglycerol (MLCT) further reduces body fat accumulation in DIO C57BL/6J mice when consumed in low fat diet for a mid-term period. *Food Res. Int.* **2018**, *103*, 200–207. [CrossRef]
72. Xue, C.; Liu, Y.; Wang, J.; Zhang, R.; Zhang, Y.; Zhang, J.; Zheng, Z.; Yu, X.; Jing, H.; Nosaka, N. Consumption of medium-and long-chain triacylglycerols decreases body fat and blood triglyceride in Chinese hypertriglyceridemic subjects. *Eur. J. Clin. Nutr.* **2009**, *63*, 879–886. [CrossRef]
73. Barness, L.A. History of infant feeding practices. *Am. J. Clin. Nutr.* **1987**, *46*, 168–170. [CrossRef]
74. Powers, G.F. Comparison and interpretation on a caloric basis of the milk mixtures used in infant feeding. *Am. J. Dis. Child.* **1925**, *30*, 453–475. [CrossRef]
75. Gerstenberger, H.J.; Ruh, H.O.; Brickman, M.J.; Leslie, H.J.; Ochsner, R.J. Studies in the Adaptation of an Artificial Food to Human Milk. *Am. J. Dis. Child.* **1915**, *10*, 249–265. [CrossRef]
76. Schuchardt, J.P.; Huss, M.; Stauss-Grabo, M.; Hahn, A. Significance of long-chain polyunsaturated fatty acids (PUFAs) for the development and behaviour of children. *Eur. J. Pediatr.* **2010**, *169*, 149–164. [CrossRef] [PubMed]
77. Bach, A.; Babayan, V. Medium-chain triglycerides: An update. *Am. J. Clin. Nutr.* **1982**, *36*, 950–962. [CrossRef]
78. Nelson, S.E.; Rogers, R.R.; Frantz, J.A.; Ziegler, E.E. Palm olein in infant formula: Absorption of fat and minerals by normal infants. *Am. J. Clin. Nutr.* **1996**, *64*, 291–296. [CrossRef]
79. Nelson, S.E.; Frantz, J.A.; Ziegler, E.E. Absorption of fat and calcium by infants fed a milk-based formula containing palm olein. *J. Am. Collage Nutr.* **1998**, *17*, 327–332. [CrossRef] [PubMed]
80. Specker, B.L.; Beck, A.; Kalkwarf, H.; Ho, M. Randomized trial of varying mineral intake on total body bone mineral accretion during the first year of life. *Pediatrics* **1997**, *99*, 12. [CrossRef]
81. Kennedy, K.; Fewtrell, M.S.; Morley, R.; Abbott, R.; Quinlan, P.T.; Wells, J.C.; Bindels, J.G.; Lucas, A. Double-blind, randomized trial of a synthetic triacylglycerol in formula-fed term infants: Effects on stool biochemistry, stool characteristics, and bone mineralization. *Am. J. Clin. Nutr.* **1999**, *70*, 920–927. [CrossRef]
82. Furse, S.; Koulman, A. The Lipid and Glyceride Profiles of Infant Formula Differ by Manufacturer, Region and Date Sold. *Nutrients* **2019**, *11*, 1122. [CrossRef]
83. Lien, E.; Richard, C.; Hoffman, D. DHA and ARA addition to infant formula: Current status and future research directions. *Prostaglandins Leukot. Essent. Fat. Acids* **2018**, *128*, 26–40. [CrossRef] [PubMed]
84. Durmuş, M. Fish oil for human health: Omega-3 fatty acid profiles of marine seafood species. *Food Sci. Technol.* **2018**, *39*, 454–461. [CrossRef]
85. Hemaiswarya, S.; Raja, R.; Kumar, R.R.; Ganesan, V.; Anbazhagan, C. Microalgae: A sustainable feed source for aquaculture. *World J. Microbiol. Biotechnol.* **2011**, *27*, 1737–1746. [CrossRef]
86. Morris, G.; Moorcraft, J.; Mountjoy, A.; Wells, J. A novel infant formula milk with added long-chain polyunsaturated fatty acids from single-cell sources: A study of growth, satisfaction and health. *Eur. J. Clin. Nutr.* **2000**, *54*, 883–886. [CrossRef]
87. Lands, B. Historical perspectives on the impact of n-3 and n-6 nutrients on health. *Prog. Lipid Res.* **2014**, *55*, 17–29. [CrossRef]
88. Colombo, J.; Kannass, K.N.; Jill Shaddy, D.; Kundurthi, S.; Maikranz, J.M.; Anderson, C.J.; Blaga, O.M.; Carlson, S.E. Maternal DHA and the development of attention in infancy and toddlerhood. *Child Dev.* **2004**, *75*, 1254–1267. [CrossRef]
89. Hsieh, A.T.; Anthony, J.C.; Diersen-Schade, D.A.; Rumsey, S.C.; Lawrence, P.; Li, C.; Nathanielsz, P.W.; Brenna, J.T. The influence of moderate and high dietary long chain polyunsaturated fatty acids (LCPUFA) on baboon neonate tissue fatty acids. *Pediatr. Res.* **2007**, *61*, 537–545. [CrossRef]
90. Mansson, H.L. Fatty acids in bovine milk fat. *Food Nutr. Res.* **2008**, *52*, 1821. [CrossRef]
91. Bhatnagar, A.S.; Prasanth Kumar, P.K.; Hemavathy, J.; Gopala Krishna, A.G. Fatty Acid Composition, Oxidative Stability, and Radical Scavenging Activity of Vegetable Oil Blends with Coconut Oil. *J. Am. Oil Chem. Soc.* **2009**, *86*, 991–999. [CrossRef]
92. Kostik, V.; Memeti, S.; Bauer, B. Fatty acid composition of edible oils and fats. *J. Hyg. Eng. Des.* **2013**, *4*, 112–116.
93. Zambiazi, R.C.; Przybylski, R.; Zambiazi, M.W.; Mendonça, C.B. Fatty acid composition of vegetable oils and fats. *Bol. Cent. Pesqui. Processamento Aliment.* **2007**, *25*, 111–120.
94. Arterburn, L.M.; Boswell, K.D.; Lawlor, T.; Cifone, M.A.; Kyle, D.J. In vitro genotoxicity testing of ARASCO and DHASCO oils. *Food Chem. Toxicol.* **2000**, *38*, 971–976. [CrossRef]
95. Guil-Guerrero, J.L.; Belarbi, E.H. Purification process for cod liver oil polyunsaturated fatty acids. *J. Am. Oil Chem. Soc.* **2001**, *78*, 477–484. [CrossRef]
96. Koriyama, T.; Wongso, S.; Watanabe, K.; Abe, H. Fatty Acid Compositions of Oil Species Affect the 5 Basic Taste Perceptions. *J. Food Sci.* **2002**, *67*, 868–873. [CrossRef]

97. Zhu, M.; Zhou, P.P.; Yu, L.J. Extraction of lipids from *Mortierella alpina* and enrichment of arachidonic acid from the fungal lipids. *Bioresour. Technol.* **2002**, *84*, 93–95. [CrossRef]
98. Wynn, J. Production of Single Cell Oils by Dinoflagellates. In *Single Cell Oils*, 2nd ed.; AOCS Press: Urbana, IL, USA, 2010; pp. 115–129.
99. Tengku-Rozaina, T.; Birch, E.J. Physicochemical characterisation and oxidative stability of refined hoki oil, unrefined hoki oil and unrefined tuna oil. *Int. J. Food Sci. Technol.* **2013**, *48*, 2331–2339. [CrossRef]
100. Sun, C.; Wei, W.; Su, H.; Zou, X.; Wang, X. Evaluation of sn-2 fatty acid composition in commercial infant formulas on the Chinese market: A comparative study based on fat source and stage. *Food Chem.* **2018**, *242*, 29–36. [CrossRef]
101. Zou, X.Q.; Huang, J.H.; Jin, Q.Z.; Liu, Y.F.; Tao, G.J.; Cheong, L.Z.; Wang, X.G. Preparation of human milk fat substitutes from palm stearin with arachidonic and docosahexaenoic acid: Combination of enzymatic and physical methods. *J. Agric. Food Chem.* **2012**, *60*, 9415–9423. [CrossRef]
102. Wang, X.; Chen, Y.; Zheng, L.; Jin, Q.; Wang, X. Synthesis of 1, 3-distearoyl-2-oleoylglycerol by enzymatic acidolysis in a solvent-free system. *Food Chem.* **2017**, *228*, 420–426. [CrossRef]
103. Tomarelli, R.M.; Meyer, B.J.; Weaber, J.R.; Bernhart, F.W. Effect of positional distribution on the absorption of the fatty acids of human milk and infant formulas. *J. Nutr.* **1968**, *95*, 583. [CrossRef]
104. Iwasaki, Y.; Yamane, T. Enzymatic synthesis of structured lipids. In *Recent Progress of Biochemical and Biomedical Engineering in Japan I.*; Springer Science & Business Media: Berlin/Heidelberg, Germany, 2004; pp. 151–171.
105. Yue, C.; Ben, H.; Wang, J.; Li, T.; Yu, G. Ultrasonic pretreatment in synthesis of caprylic-rich structured lipids by lipase-catalyzed acidolysis of corn oil in organic system and its physicochemical properties. *Foods* **2019**, *8*, 566. [CrossRef] [PubMed]
106. Chang, C.D.; Waki, M.; Ahmad, M.; Meienhofer, J.; Lundell, E.O.; Haug, J.D. Preparation and properties of nα-9-fluorenylmethyloxycarbonylamino acids bearing tert.-butyl side chain protection. *Int. J. Pept. Protein Res.* **1980**, *15*, 59–66. [CrossRef] [PubMed]
107. Wei, W.; Sun, C.; Wang, X.; Jin, Q.; Xu, X.; Akoh, C.C.; Wang, X. Lipase-Catalyzed Synthesis of Sn-2 Palmitate: A Review—ScienceDirect. *Engineering* **2020**, *6*, 406–414. [CrossRef]
108. Ilyasoglu, H.; Gultekin-Ozguven, M.; Ozcelik, B. Production of human milk fat substitute with medium-chain fatty acids by lipase-catalyzed acidolysis: Optimization by response surface methodology. *LWT-Food Sci. Technol.* **2011**, *44*, 999–1004. [CrossRef]
109. Suxi, W.; Min, J.; Meng, G.E. Comparison of fatty acid composition, structure and functional properties of palm oil and lard. *China Oils Fats* **2009**, *34*, 39–44.
110. Yang, T.; Xu, X.; He, C.; Li, L. Lipase-catalyzed modification of lard to produce human milk fat substitutes. *Food Chem.* **2003**, *80*, 473–481. [CrossRef]
111. Zhao, H.; Lu, Z.; Lu, F.; Bie, X.; Liu, Z.; Zeng, X. Lipase-catalysed acidolysis of lard with caprylic acid to produce structured lipid. *Int. J. Food Sci. Technol.* **2006**, *41*, 1027–1032. [CrossRef]
112. Nielsen, N.S.; Yang, T.; Xu, X.; Jacobsen, C. Production and oxidative stability of a human milk fat substitute produced from lard by enzyme technology in a pilot packed-bed reactor. *Food Chem.* **2006**, *94*, 53–60. [CrossRef]
113. Zou, X.; Jin, Q.; Guo, Z.; Huang, J.; Xu, X.; Wang, X. Preparation of 1, 3-dioleoyl-2-palmitoylglycerol-rich structured lipids from basa catfish oil: Combination of fractionation and enzymatic acidolysis. *Eur. J. Lipid Sci. Technol.* **2016**, *118*, 708–715. [CrossRef]
114. He, Y.; Qiu, C.; Guo, Z.; Huang, J.; Wang, M.; Chen, B. Production of new human milk fat substitutes by enzymatic acidolysis of microalgae oils from *Nannochloropsis oculata* and *Isochrysis galbana*. *Bioresour. Technol.* **2017**, *238*, 129–138. [CrossRef]
115. Jiménez, M.J.; Esteban, L.; Robles, A.; Hita, E.; González, P.A.; Muñío, M.M.; Molina, E. Production of triacylglycerols rich in palmitic acid at sn-2 position by lipase-catalyzed acidolysis. *Biochem. Eng. J.* **2010**, *51*, 172–179. [CrossRef]
116. Jiménez, M.J.; Esteban, L.; Robles, A.; Hita, E.; González, P.A.; Muñío, M.M.; Molina, E. Production of triacylglycerols rich in palmitic acid at position 2 as intermediates for the synthesis of human milk fat substitutes by enzymatic acidolysis. *Process Biochem.* **2010**, *45*, 407–414. [CrossRef]
117. Esteban, L.; Jiménez, M.J.; Hita, E.; González, P.A.; Martín, L.; Robles, A. Production of structured triacylglycerols rich in palmitic acid at sn-2 position and oleic acid at sn-1,3 positions as human milk fat substitutes by enzymatic acidolysis. *Biochem. Eng. J.* **2011**, *54*, 62–69. [CrossRef]
118. Sahin, N.; Akoh, C.C.; Karaali, A. Lipase-catalyzed acidolysis of tripalmitin with hazelnut oil fatty acids and stearic acid to produce human milk fat substitutes. *J. Agric. Food Chem.* **2005**, *53*, 5779–5783. [CrossRef] [PubMed]
119. Li, R.; Pande, G.; Sabir, J.S.; Baeshen, N.A.; Akoh, C.C. Enrichment of refined olive oil with palmitic and docosahexaenoic acids to produce a human milk fat analogue. *J. Am. Oil Chem. Soc.* **2014**, *91*, 1377–1385. [CrossRef]
120. Pande, G.; Sabir, J.S.; Baeshen, N.A.; Akoh, C.C. Synthesis of infant formula fat analogs enriched with DHA from extra virgin olive oil and tripalmitin. *J. Am. Oil Chem. Soc.* **2013**, *90*, 1311–1318. [CrossRef]
121. Nagachinta, S.; Akoh, C.C. Production and Characterization of DHA and GLA-Enriched Structured Lipid from Palm Olein for Infant Formula Use. *J. Am. Oil Chem. Soc.* **2013**, *90*, 1141–1149. [CrossRef]
122. Nagachinta, S.; Akoh, C.C. Synthesis of structured lipid enriched with omega fatty acids and sn-2 palmitic acid by enzymatic esterification and its incorporation in powdered infant formula. *J. Agric. Food Chem.* **2013**, *61*, 4455–4463. [CrossRef]
123. Wan, J.; Hu, S.; Jacoby, J.J.; Liu, J.; Zhang, Y.; Yu, L.L. The impact of dietary sn-2 palmitic triacylglycerols in combination with docosahexaenoic acid or arachidonic acid on lipid metabolism and host faecal microbiota composition in Sprague Dawley rats. *Food Funct.* **2017**, *8*, 1793–1802. [CrossRef]

124. Zou, X.; Ye, L.; He, X.; Wu, S.; Zhang, H.; Jin, Q. Preparation of DHA-rich medium-and long-chain triacylglycerols by lipase-catalyzed acidolysis of microbial oil from *Schizochytrium* sp. with medium-chain fatty acids. *Appl. Biochem. Biotechnol.* **2020**, *191*, 1294–1314. [CrossRef]
125. Naranjo, J.M.D.; Callejón, M.J.J.; Vásquez, M.P.; Rios, L.A.; Medina, A.R. Optimization of the enzymatic synthesis of structured triacylglycerols rich in docosahexaenoic acid at sn-2 position by acidolysis of Aurantiochytrium limacinum SR21 oil and caprylic acid using response surface methodology. *J. Appl. Phycol.* **2021**, *33*, 2031–2045. [CrossRef]
126. Maduko, C.; Park, Y. Modification of fatty acid and sterol composition of caprine milk for use as infant formula. *Int. Dairy J.* **2007**, *17*, 1434–1440. [CrossRef]
127. Korma, S.A.; Zou, X.; Ali, A.H.; Abed, S.M.; Jin, Q.; Wang, X. Preparation of structured lipids enriched with medium-and long-chain triacylglycerols by enzymatic interesterification for infant formula. *Food Bioprod. Process* **2018**, *107*, 121–130. [CrossRef]
128. Ghosh, M.; Sengupta, A.; Bhattacharyya, D.; Ghosh, M. Preparation of human milk fat analogue by enzymatic interesterification reaction using palm stearin and fish oil. *J. Food Sci. Technol.* **2016**, *53*, 2017–2024. [CrossRef] [PubMed]
129. Bektaş, I.; Yucel, S.; Ustun, G.; Aksoy, H.A. Production of reduced calorie structured lipid by acidolysis of tripalmitin with capric acid: Optimisation by response surface methodology. *J. Sci. Food Agric.* **2008**, *88*, 1927–1931. [CrossRef]
130. Pina-Rodriguez, A.M.; Akoh, C.C. Synthesis and characterization of a structured lipid from amaranth oil as a partial fat substitute in milk-based infant formula. *J. Agric. Food Chem.* **2009**, *57*, 6748–6756. [CrossRef]
131. Aresta, M.; Dibenedetto, A.; Nocito, F.; Ferragina, C. Valorization of bio-glycerol: New catalytic materials for the synthesis of glycerol carbonate via glycerolysis of urea. *J. Catal.* **2009**, *268*, 106–114. [CrossRef]
132. Duan, Z.-Q.; Du, W.; Liu, D.-H. Novozym 435-catalyzed 1, 3-diacylglycerol preparation via esterification in t-butanol system. *Process Biochem.* **2010**, *45*, 1923–1927. [CrossRef]
133. Yang, K.; Bi, Y.; Sun, S.; Yang, G.; Ma, S.; Liu, W. Optimisation of N ovozym-435-catalysed esterification of fatty acid mixture for the preparation of medium-and long-chain triglycerides (MLCT) in solvent-free medium. *Int. J. Food Sci. Technol.* **2014**, *49*, 1001–1011. [CrossRef]
134. Nagao, T.; Watanabe, Y.; Maruyama, K.; Momokawa, Y.; Kishimoto, N.; Shimada, Y. One-pot enzymatic synthesis of docosahexaenoic acid-rich triacylglycerols at the sn-1(3) position using by-product from selective hydrolysis of tuna oil. *New Biotechnol.* **2011**, *28*, 7–13. [CrossRef]
135. Agapay, R.C.; Ju, Y.H.; Tran-Nguyen, P.L.; Ismadji, S.; Angkawijaya, A.E.; Go, A.W. Process evaluation of solvent-free lipase-catalyzed esterification schemes in the synthesis of structured triglycerides from oleic and palmitic acids. *Asia-Pac. J. Chem. Eng.* **2020**, *16*, 2606. [CrossRef]
136. Morales-Medina, R.; Munio, M.; Guadix, A.; Guadix, E. Development of an up-grading process to produce MLM structured lipids from sardine discards. *Food Chem.* **2017**, *228*, 634–642. [CrossRef] [PubMed]
137. Schmid, U.; Bornscheuer, U.; Soumanou, M.; McNeill, G.; Schmid, R. Highly selective synthesis of 1, 3-oleoyl-2-palmitoylglycerol by lipase catalysis. *Biotechnol. Bioeng.* **1999**, *64*, 678–684. [CrossRef]
138. Pfeffer, J.; Freund, A.; Bel-Rhlid, R.; Hansen, C.E.; Reuss, M.; Schmid, R.D.; Maurer, S.C. Highly efficient enzymatic synthesis of 2-monoacylglycerides and structured lipids and their production on a technical scale. *Lipids* **2007**, *42*, 947–953. [CrossRef]
139. Irimescu, R.; Furihata, K.; Hata, K.; Iwasaki, Y.; Yamane, T. Utilization of reaction medium-dependent regiospecificity of Candida antarctica lipase (Novozym 435) for the synthesis of 1,3-dicapryloyl-2-docosahexaenoyl (or eicosapentaenoyl) glycerol. *J. Am. Oil Chem. Soc.* **2001**, *78*, 285–290. [CrossRef]
140. Tecelão, C.; Guillén, M.; Valero, F.; Ferreira-Dias, S. Immobilized heterologous Rhizopus oryzae lipase: A feasible biocatalyst for the production of human milk fat substitutes. *Biochem. Eng. J.* **2012**, *67*, 104–110. [CrossRef]
141. Wei, W.; Feng, Y.; Zhang, X.; Cao, X.; Feng, F. Synthesis of structured lipid 1,3-dioleoyl-2-palmitoylglycerol in both solvent and solvent-free system. *LWT-Food Sci. Technol.* **2015**, *60*, 1187–1194. [CrossRef]
142. Zheng, M.; Wang, S.; Xiang, X.; Shi, J.; Huang, J.; Deng, Q.; Huang, F.; Xiao, J. Facile preparation of magnetic carbon nanotubes-immobilized lipase for highly efficient synthesis of 1,3-dioleoyl-2-palmitoylglycerol-rich human milk fat substitutes. *Food Chem.* **2017**, *228*, 476–483. [CrossRef]
143. Can, A.; Özçelik, B. Enrichment of hazelnut oil with long-chain n− 3 PUFA by lipase-catalyzed acidolysis: Optimization by response surface methodology. *J. Am. Oil Chem. Soc.* **2005**, *82*, 27–32. [CrossRef]
144. Abed, S.M.; Zou, X.; Ali, A.H.; Jin, Q.; Wang, X. Synthesis of 1,3-dioleoyl-2-arachidonoylglycerol-rich structured lipids by lipase-catalyzed acidolysis of microbial oil from *Mortierella alpina*. *Bioresour. Technol.* **2017**, *243*, 448–456. [CrossRef]
145. Hamam, F.; Shahidi, F. Enzymatic incorporation of capric acid into a single cell oil rich in docosahexaenoic acid and docosapentaenoic acid and oxidative stability of the resultant structured lipid. *Food Chem.* **2005**, *91*, 583–591. [CrossRef]
146. Nagao, T.; Kawashima, A.; Sumida, M.; Watanabe, Y.; Akimoto, K.; Fukami, H.; Sugihara, A.; Shimada, Y. Production of structured TAG rich in 1, 3-capryloyl-2-arachidonoyl glycerol from Mortierella single-cell oil. *J. Am. Oil Chem. Soc.* **2003**, *80*, 867–872. [CrossRef]
147. Wang, X.; Zou, S.; Miu, Z.; Jin, Q.; Wang, X. Enzymatic preparation of structured triacylglycerols with arachidonic and palmitic acids at the sn-2 position for infant formula use. *Food Chem.* **2019**, *283*, 331–337. [CrossRef] [PubMed]
148. Lee, N.K.; Oh, S.W.; Kwon, D.Y.; Yoon, S.H. Production of 1, 3-dioleoyl-2-palmitoyl glycerol as a human milk fat substitute using enzymatic interesterification of natural fats and oils. *Food Sci. Biotechnol.* **2015**, *24*, 433–437. [CrossRef]

149. Chen, M.-L.; Vali, S.R.; Lin, J.-Y.; Ju, Y.-H. Synthesis of the structured lipid 1,3-dioleoyl-2-palmitoylglycerol from palm oil. *J. Am. Oil Chem. Soc.* **2004**, *81*, 525–532. [CrossRef]
150. Lee, J.H.; Son, J.M.; Akoh, C.; Kim, M.R.; Lee, K.T. Optimized synthesis of 1,3-dioleoyl-2-palmitoylglycerol-rich triacylglycerol via interesterification catalyzed by a lipase from Thermomyces lanuginosus. *New Biotechnol.* **2010**, *27*, 38–45. [CrossRef] [PubMed]
151. Irimescu, R.; Furihata, K.; Hata, K.; Iwasaki, Y.; Yamane, T. Two-step enzymatic synthesis of docosahexaenoic acid-rich symmetrically structured triacylglycerols via 2-monoacylglycerols. *J. Am. Oil Chem. Soc.* **2001**, *78*, 743–748. [CrossRef]
152. Zhang, L.-S.; Chu, M.-Y.; Zong, M.-H.; Yang, J.-G.; Lou, W.-Y. Carbon source modify lipids composition of *Rhodococcus opacus* intended for infant formula. *J. Biotechnol.* **2020**, *319*, 8–14. [CrossRef]
153. Zhang, L.S.; Chu, M.Y.; Zong, M.H.; Yang, J.G.; Lou, W.Y. Efficient Production of 1, 3-Dioleoyl-2-Palmitoylglycerol through *Rhodococcus opacus* Fermentation. *J. Am. Oil Chem. Soc.* **2020**, *97*, 851–860. [CrossRef]
154. Gupta, R.; Rathi, P.; Gupta, N.; Bradoo, S. Lipase assays for conventional and molecular screening: An overview. *Biotechnol. Appl. Biochem.* **2003**, *37*, 63–71. [CrossRef]
155. Huang, X.-J.; Chen, P.-C.; Huang, F.; Ou, Y.; Chen, M.-R.; Xu, Z.-K. Immobilization of Candida rugosa lipase on electrospun cellulose nanofiber membrane. *J. Mol. Catal. B Enzym.* **2011**, *70*, 95–100. [CrossRef]
156. Park, Y.-K.; Dulermo, T.; Ledesma-Amaro, R.; Nicaud, J.-M. Optimization of odd chain fatty acid production by *Yarrowia lipolytica*. *Biotechnol. Biofuels* **2018**, *11*, 158. [CrossRef] [PubMed]
157. de Moura, J.M.; Gonçalves, L.A.; Sarmento, L.A.; Petrus, J.C.C. Purification of structured lipids using SCCO2 and membrane process. *J. Membr. Sci.* **2007**, *299*, 138–145. [CrossRef]
158. Ji, S.; Wu, J.; Xu, F.; Wu, Y.; Xu, X.; Gao, H.; Ju, X.; Xiong, W.; Wang, L. Synthesis, Purification, and Characterization of a Structured Lipid Based on Soybean Oil and Coconut Oil and Its Applications in Curcumin-Loaded Nanoemulsions. *Eur. J. Lipid Sci. Technol.* **2020**, *122*, 2000086. [CrossRef]
159. Catchpole, O.; Tallon, S.; Eltringham, W.; Grey, J.; Fenton, K.; Vagi, E.; Vyssotski, M.; MacKenzie, A.; Ryan, J.; Zhu, Y. The extraction and fractionation of specialty lipids using near critical fluids. *J. Supercrit. Fluids* **2009**, *47*, 591–597. [CrossRef]
160. Čmolík, J.; Pokorný, J. Physical refining of edible oils. *Eur. J. Lipid Sci. Technol.* **2000**, *102*, 472–486. [CrossRef]
161. Tao, L. Oxidation of polyunsaturated fatty acids and its impact on food quality and human health. *Adv. Food Technol. Nutr. Sci.* **2015**, *1*, 135–142. [CrossRef]
162. Chen, W.; Wang, H.; Zhang, K.; Gao, F.; Chen, S.; Li, D. Physicochemical properties and storage stability of microencapsulated DHA-rich oil with different wall materials. *Appl. Biochem. Biotechnol.* **2016**, *179*, 1129–1142. [CrossRef]
163. Loughrill, E.; Thompson, S.; Owusu-Ware, S.; Snowden, M.J.; Douroumis, D.; Zand, N. Controlled release of microencapsulated docosahexaenoic acid (DHA) by spray–drying processing. *Food Chem.* **2019**, *286*, 368–375. [CrossRef]
164. Karthik, P.; Anandharamakrishnan, C. Microencapsulation of docosahexaenoic acid by spray-freeze-drying method and comparison of its stability with spray-drying and freeze-drying methods. *Food Bioprocess Technol.* **2013**, *6*, 2780–2790. [CrossRef]
165. Ré, M.I. Microencapsulation by spray drying. *Dry. Technol.* **1998**, *16*, 1195–1236. [CrossRef]
166. Patel, R.; Patel, M.; Suthar, A. Spray drying technology: An overview. *Indian J. Sci. Technol.* **2009**, *2*, 44–47. [CrossRef]
167. Vélez-Erazo, E.M.; Consoli, L.; Hubinger, M.D. Spray drying of mono-and double-layer emulsions of PUFA-rich vegetable oil homogenized by ultrasound. *Dry. Technol.* **2021**, *39*, 868–881. [CrossRef]
168. Santos, D.; Maurício, A.C.; Sencadas, V.; Santos, J.D.; Fernandes, M.H.; Gomes, P.S. *Spray Drying: An Overview*; IntechOpen: London, UK, 2018; pp. 9–35.
169. El-Loly, M.M. Composition, properties and nutritional aspects of milk fat globule membrane—A review. *Pol. J. Food Nutr. Sci.* **2011**, *61*, 7–32. [CrossRef]
170. Anderson, M. Milk fat globule membrane composition and dietary change: Supplements of coconut oil fed in two physical forms. *J. Dairy Sci.* **1974**, *57*, 399–404. [CrossRef]
171. Erickson, D.; Dunkley, W.; Smith, L. Tocopherol Distribution in Milk Fractions and Its Relation to Antioxidant Activity. *J. Food Sci.* **1964**, *29*, 269–275. [CrossRef]
172. Nejjar, Y.; Paquet, D.; Aubert, F.; Linden, G. The PP3 component of the proteose-peptone. Extraction from unheated skim milk. *Milchwissenschaft* **1990**, *45*, 84–87.
173. Morin, P.; Britten, M.; Jiménez-Flores, R.; Pouliot, Y. Microfiltration of buttermilk and washed cream buttermilk for concentration of milk fat globule membrane components. *J. Dairy Sci.* **2007**, *90*, 2132–2140. [CrossRef]
174. Morin, P.; Jiménez-Flores, R.; Pouliot, Y. Effect of temperature and pore size on the fractionation of fresh and reconstituted buttermilk by microfiltration. *J. Dairy Sci.* **2004**, *87*, 267–273. [CrossRef]
175. Rombaut, R.; Dewettinck, K. Properties, analysis and purification of milk polar lipids. *Int. Dairy J.* **2006**, *16*, 1362–1373. [CrossRef]
176. Akoh, C.C.; Xu, X. Enzymatic production of Betapol and other specialty fats. In *Lipid Biotechnology*; CRC Press: Boca Raton, FL, USA, 2002; pp. 534–554.
177. Yang, P.; Zhang, H.; Wan, J.; Hu, J.; Liu, J.; Wang, J.; Zhang, Y.; Yu, L. Dietary sn-2 palmitic triacylglycerols reduced faecal lipids, calcium contents and altered lipid metabolism in Sprague–Dawley rats. *Int. J. Food Sci. Nutr.* **2019**, *70*, 474–483. [CrossRef] [PubMed]
178. Ghide, M.K.; Yan, Y. 1,3-Dioleoyl-2-palmitoyl glycerol (OPO)-Enzymatic synthesis and use as an important supplement in infant formulas. *J. Food Biochem.* **2021**, *45*, e13799. [CrossRef] [PubMed]

179. Schmid, U.; Bornscheuer, U.; Soumanou, M.; McNeill, G.; Schmid, R. Optimization of the reaction conditions in the lipase-catalyzed synthesis of structured triglycerides. *J. Am. Oil Chem. Soc.* **1998**, *75*, 1527–1531. [CrossRef]
180. Timby, N.; Domellöf, M.; Lönnerdal, B.; Hernell, O. Supplementation of infant formula with bovine milk fat globule membranes. *Adv. Nutr.* **2017**, *8*, 351–355. [CrossRef] [PubMed]
181. Hernell, O.; Timby, N.; Domellöf, M.; Lönnerdal, B. Clinical benefits of milk fat globule membranes for infants and children. *J. Pediatr.* **2016**, *173*, S60–S65. [CrossRef]
182. Korff, F.A. Federal Food, Drug and Cosmetic Act, Judicial and Administrative Record, 1938–1949. *Am. J. Public Health Nations Health* **1950**, *40*, 885–886. [CrossRef]
183. Pilot, L.R. Federal food, drug, and cosmetic act. In *Pharmacy Law Desk Reference*; Routledge: London, UK, 2012; pp. 43–57.
184. Walsh, G.P. *Federal Food, Drug, and Cosmetic Act with Amendments*; US Government Printing Office: Washington, DC, USA, 1981.
185. Code pf Federal Regulations. Available online: https://www.ecfr.gov/current/title-21/chapter-I/subchapter-B/part-107/subpart-D/section-107.100 (accessed on 31 November 2021).
186. Commission Delegated Regulation (EU) 2016/127 of 25 September 2015 Supplementing Regulation (EU) No. 609/2013 of the European Parliament and of the Council as Regards the Specific Compositional and Information Requirements for Infant Formula and Follow-On Formula and as Regards Requirements on Information Relating to Infant and Young Child Feeding (Text with EEA Relevance). Available online: https://eur-lex.europa.eu/legal-content/EN/TXT/?uri=CELEX%3A32016R0127&qid=1643206644591 (accessed on 31 November 2021).
187. Koletzko, S.; Niggemann, B.; Arato, A.; Dias, J.; Heuschkel, R.; Husby, S.; Mearin, M.; Papadopoulou, A.; Ruemmele, F.; Staiano, A. European Society of Pediatric Gastroenterology, Hepatology, and Nutrition. Diagnostic approach and management of cow's-milk protein allergy in infants and children: ESPGHAN GI Committee practical guidelines. *J. Pediatr. Gastroenterol. Nutr.* **2012**, *55*, 221–229. [CrossRef]
188. Klein, C.J. Nutrient requirements for preterm infant formulas. *J. Nutr.* **2002**, *132*, 1395–1577. [CrossRef]
189. Alimentarius, Codex. *Joint FAO/WHO Food Standards Programme*; Codex Alimentarius Commission, FAO: Rome, Italy, 1994; Volume 11, p. 22.
190. Raiten, D.J.; Talbot, J.M.; Waters, J.H. LSRO report: Assessment of nutrient requirements for infant formulas-Foreword. *J. Nutr.* **1998**, *128*, 2059S–2293S.
191. Report of the Scientific Committee on Food on the Revision of Essential Requirements of Infant Formulae and Follow-On Formulae. Available online: https://ec.europa.eu/food/system/files/2020-12/sci-com_scf_out199_en.pdf (accessed on 31 November 2021).
192. Yates, A.A.; Schlicker, S.A.; Suitor, C.W. Dietary reference intakes: The new basis for recommendations for calcium and related nutrients, B vitamins, and choline. *J. Am. Diet Assoc.* **1998**, *98*, 699–706. [CrossRef]
193. Cantellops, D.; Reid, A.P.; Eitenmiller, R.R.; Long, A.R. Determination of lipids in infant formula powder by direct extraction methylation of lipids and fatty acid methyl esters (FAME) analysis by gas chromatography. *J. AOAC Int.* **1999**, *82*, 1128–1139. [CrossRef] [PubMed]
194. Fong, B.; Ma, L.; Norris, C. Analysis of phospholipids in infant formulas using high performance liquid chromatography–tandem mass spectrometry. *J. Agric. Food Chem.* **2013**, *61*, 858–865. [CrossRef] [PubMed]
195. Bryś, J.; Wirkowska, M.; Górska, A.; Ostrowska-Ligęza, E.; Bryś, A. Application of the calorimetric and spectroscopic methods in analytical evaluation of the human milk fat substitutes. *J. Therm. Anal. Calorim.* **2014**, *118*, 841–848. [CrossRef]
196. Zou, X.-G.; Hu, J.-N.; Zhao, M.-L.; Zhu, X.-M.; Li, H.-Y.; Liu, X.-R.; Liu, R.; Deng, Z.-Y. Lipozyme RM IM-catalyzed acidolysis of *Cinnamomum camphora* seed oil with oleic acid to produce human milk fat substitutes enriched in medium-chain fatty acids. *J. Agric. Food Chem.* **2014**, *62*, 10594–10603. [CrossRef] [PubMed]
197. Xue, L.; Cao, J.; Bai, X.-P.; Zhang, F. Monitoring of Oxidation Process of Human Milk Lipid Substitutes by Molecular Fluorescence Spectroscopy. *Chin. J. Anal. Chem.* **2018**, *46*, 543–549.
198. Zou, X.Q.; Huang, J.H.; Jin, Q.Z.; Guo, Z.; Liu, Y.F.; Cheong, L.Z.; Xu, X.B.; Wang, X.G. Model for human milk fat substitute evaluation based on triacylglycerol composition profile. *J. Agric. Food Chem.* **2013**, *61*, 167–175. [CrossRef] [PubMed]
199. Kloek, W.; Vonk, M.M.; Feitsma, A.L.; Timmer, C.J. Application of the similarity index to evaluate fat composition and structure in infant formulas. *Int. Dairy J.* **2020**, *111*, 104834. [CrossRef]
200. Zou, X.; Jin, Q.; Guo, Z.; Xu, X.; Wang, X. Preparation of human milk fat substitutes from basa catfish oil: Combination of enzymatic acidolysis and modeled blending. *Eur. J. Lipid Sci. Technol.* **2016**, *118*, 1702–1711. [CrossRef]
201. Wang, Y.H.; Qin, X.L.; Zhu, Q.S.; Zhou, R.; Yang, B.; Li, L. Lipase-catalyzed acidolysis of lard for the production of human milk fat substitute. *Eur. Food Res. Technol.* **2010**, *230*, 769–777. [CrossRef]
202. Zou, X.-Q.; Huang, J.-H.; Jin, Q.-Z.; Liu, Y.-F.; Song, Z.-H.; Wang, X.-G. Lipase-catalyzed preparation of human milk fat substitutes from palm stearin in a solvent-free system. *J. Agric. Food Chem.* **2011**, *59*, 6055–6063. [CrossRef]
203. Zou, X.; Huang, J.; Jin, Q.; Liu, Y.; Song, Z.; Wang, X. Lipase-catalyzed synthesis of human milk fat substitutes from palm stearin in a continuous packed bed reactor. *J. Am. Oil Chem. Soc.* **2012**, *89*, 1463–1472. [CrossRef]
204. Wang, X.; Huang, Z.; Hua, L.; Zou, F.; Cheng, X.; Wang, X. Preparation of human milk fat substitutes similar to human milk fat by enzymatic acidolysis and physical blending. *LWT* **2021**, *140*, 110818. [CrossRef]

205. Zou, X.; Huang, J.; Jin, Q.; Guo, Z.; Liu, Y.; Cheong, L.; Xu, X.; Wang, X. Lipid composition analysis of milk fats from different mammalian species: Potential for use as human milk fat substitutes. *J. Agric. Food Chem.* **2013**, *61*, 7070–7080. [CrossRef] [PubMed]
206. Sun, C.; Wei, W.; Zou, X.; Huang, J.; Jin, Q.; Wang, X. Evaluation of triacylglycerol composition in commercial infant formulas on the Chinese market: A comparative study based on fat source and stage. *Food Chem.* **2018**, *252*, 154–162. [CrossRef]

MDPI
St. Alban-Anlage 66
4052 Basel
Switzerland
Tel. +41 61 683 77 34
Fax +41 61 302 89 18
www.mdpi.com

Life Editorial Office
E-mail: life@mdpi.com
www.mdpi.com/journal/life

www.ingramcontent.com/pod-product-compliance
Lightning Source LLC
Chambersburg PA
CBHW062104220926
51850CB00010B/2885